DEVELOPING AND MANAGING ENGINEERING PROCEDURES

Concepts and Applications

DEVELOPING AND MANAGING ENGINEERING PROCEDURES

Concepts and Applications

by

Phillip A. Cloud

Geneva Pharmaceutical, Inc.
Broomfield, Colorado

NOYES PUBLICATIONS
Park Ridge, New Jersey, U.S.A.

WILLIAM ANDREW PUBLISHING, LLC
Norwich, New York, U.S.A.

Published in the United States of America by
Noyes Publications / William Andrew Publishing, LLC
13 Eaton Avenue
Norwich, NY 13815
1-800-932-7045
www.knovel.com

Transferred to Digital Printing, 2010

Printed and bound in the United Kingdom

Cloud, Phillip K.
    Developing and Managing Engineering Procedures.

NOTICE

To the best of our knowledge the information in this publication is
accurate; however the Publisher does not assume any responsibility
or liability for the accuracy or completeness of, or consequences
arising from, such information. This book is intended for informational
purposes only. Mention of trade names or commercial products does
not constitute endorsement or recommendation for use by the Publisher. Final determination of the suitability of any information or
product for use contemplated by any user, and the manner of that
use, is the sole responsibility of the user. We recommend that
anyone intending to rely on any recommendation of materials or
procedures mentioned in this publication should satisfy himself as
to such suitability, and that he can meet all applicable safety and
health standards.

Preface

Industry regulations such as ISO 9000 require that engineering procedures be established and followed. These regulations do not provide guidelines on how to specifically produce them. It is left to individual companies to develop their own engineering procedures. This book details a method for writing policies, departmental instructions, and engineering procedures from the first conception to distribution for both printed and electronic media. This book explains how to develop and manage engineering procedures and their associated documentation. The book is for everyone from beginners to advanced engineering documentation practitioners.

This book explains:

- How to develop a procedure writing group
- How to determine which documents need to be written
- How to write policies, departmental instructions, and engineering procedures
- How to develop document formats and contents
- How to develop a document numbering system
- How to assign identification numbers
- How to establish a Document Review Board
- How to establish a change control system
- How to establish a document control system
- How to develop document number assignment logs

- How to publish engineering procedure manuals
- How to distribute manuals
- How documents relate to each other and to an overall documentation system

The book also provides hands-on techniques for writing engineering procedures to achieve ISO 9000 compliance. This book is designed for individuals responsible for writing engineering procedures in any industry. The book provides actual examples of engineering procedures, thus providing answers to many of your procedure-writing questions. Readers will have a general understanding of engineering documentation principles and how to apply these principles to their own industry. Simple diagrams and other graphics illustrate key ideas.

The intent of this book is to familiarize the reader with the essential elements and concepts of engineering procedure development and management and how to apply these concepts to their own specific application. Emphasis will be placed on engineering procedure principals and tools that are common to all engineering disciplines, with examples for their use. Step-by-step procedures are shown for each document format that will enable the readers to apply these approaches to their own engineering documentation programs.

This book is intended for those who are developing an engineering procedure program. It is also recommended for directors, managers and supervisors of the engineering documentation staff. This book is intended for personnel responsible for, or involved in, any element of engineering documentation development, as well as individuals interested in improving the efficiency of their own documentation efforts, including personnel in research and development, quality, and manufacturing.

This book was developed as a companion to the *Engineering Procedures Handbook* by Phil Cloud, published by William Andrews Publications.

The author's personal experience with various regulators such as FAA, NASA, EF&T, and FDA influenced the decision to write this book. The need came from always wanting to know exactly how to do something and the regulators not providing the information. The methods described in this book are those of the author and are not to be construed as the policy of his employer.

Westminster, Colorado   Phillip K. Cloud
June, 2000

Contents

1 Introduction 1

   1.1.0   ENGINEERING PROCEDURES CASE STUDY 4

   1.2.0   ENGINEERING PROCEDURES MANAGEMENT 5

   1.3.0   DOCUMENT CONTROL SYSTEM 6

   1.4.0   DOCUMENT CONTROL COMPONENTS 6

   1.5.0   WHAT EXACTLY DO YOU NEED TO DOCUMENT?.... 7

   1.6.0   ENGINEERING PROCEDURE WRITING 8

   1.7.0   HOW TO USE THIS BOOK 9

2 Engineering Procedure Writing Group 10

   2.1.0   HOW TO ESTABLISH AN ENGINEERING
PROCEDURE WRITING GROUP 11

       2.1.1   Kick-off Meeting Memo 13

       2.1.2   Agenda for the Kick-off Meeting 13

   2.2.0   BASIC WRITING GROUP FUNCTIONS 14

       2.2.1   Which Documents are Necessary? 14

       2.2.2   Establish Document Format and Contents 15

   2.3.0   BEFORE YOU START WRITING 16

       2.3.1   Interviewing 16

       2.3.2   Questions to ask for a New Procedure 16

       2.3.3   Questions to ask for a New Form........................... 17

2.4.0   POLICY WRITING TIPS 17
        2.4.1   How Many Policies are Really Needed? 18
2.5.0   DEPARTMENTAL INSTRUCTION WRITING TIPS 20
        2.5.1   How Many Departmental Instructions are Really
                Needed? 20
2.6.0   ENGINEERING PROCEDURE WRITING TIPS 20
        2.6.1   Definitions 21

3   Document Format and Contents 23
3.1.0   POLICY FORMAT 23
3.2.0   POLICY CONTENT 27
3.3.0   EXAMPLE POLICY 27
3.4.0   DEPARTMENTAL INSTRUCTION FORMAT 30
3.5.0   DEPARTMENTAL INSTRUCTION CONTENTS 34
3.6.0   EXAMPLE DEPARTMENTAL INSTRUCTION 34
3.7.0   ENGINEERING PROCEDURE FORMAT 41
3.8.0   ENGINEERING PROCEDURE CONTENTS 45
3.9.0   EXAMPLE ENGINEERING PROCEDURE 46
3.10.0 FORMS 54
3.11.0 FORM FORMAT 55
3.12.0 EXAMPLE FORM 55

4   Engineering Procedure Manuals 59
4.1.0   MANUAL COMPONENTS 60
        4.1.1   Cover Page 61
        4.1.2   Spine 62
        4.1.3   Binders 63
        4.1.4   Dividers 63
4.2.0   ENGINEERING PROCEDURE MANUAL PAGE 64
        4.2.1   Manual Introduction and Maintenance Page 66
        4.2.2   Table of Contents with Revision History Page 68
4.3.0   ENGINEERING PROCEDURE MANUAL
        DISTRIBUTION 71
        4.3.1   Distribution of Manuals to Individuals 72
        4.3.2   Distribution of Manuals to Departments 72
        4.3.3   Document Agreement Form 72
        4.3.4   Engineering Procedure Manual Distribution List . 76
        4.3.5   Example—Engineering Procedure Manual
                Distribution List 78
4.4.0   MANUAL REVISIONS 79

5   Change Control **80**

　5.1.0   DOCUMENT REVIEW AND APPROVAL 81
　　5.1.1   Engineering Department Review 82
　　5.1.2   Outside Department Review and Approval 82
　　5.1.3   Document Review Form 83
　　5.1.4   Example Document Review Form 86
　5.2.0   CHANGE CONTROL METHODS 87
　　5.2.1   Document Change Request Form 88
　　5.2.2   Example—Document Change Request Form 91
　5.3.0   DOCUMENT REVIEW BOARD 92
　5.4.0   REVISION HISTORY SHEET 93
　5.5.0   REVISION HISTORY LIST 96
　　5.5.1   Example—Revision History List 99

6   Engineering Document Control **101**

　6.1.0   BLOCK NUMBERING SYSTEM 101
　6.2.0   POLICY NUMBERS 103
　　6.2.1   Example—Policy Numbers and Interpretation 104
　6.3.0   POLICY NUMBER ASSIGNMENT 104
　　6.3.1   Policy Number Assignment Log 104
　　6.3.2   Example—Policy Number Assignment Log 106
　6.4.0   POLICY TITLES 106
　6.5.0   DEPARTMENTAL INSTRUCTION NUMBERS 108
　　6.5.1   Example—Departmental Instruction Numbers
　　　　　and Interpretation 109
　6.6.0   DEPARTMENTAL INSTRUCTIONS NUMBER
　　　　ASSIGNMENT 109
　　6.6.1   Departmental Instruction Number
　　　　　Assignment Log 109
　　6.6.2   Example—Departmental Instruction Number
　　　　　Assignment Log 111
　6.7.0   DEPARTMENTAL INSTRUCTION TITLES 111
　6.8.0   ENGINEERING PROCEDURE NUMBERS 113
　　6.8.1   Example—Engineering Procedure Numbers
　　　　　and Interpretation 114
　6.9.0   ENGINEERING PROCEDURE NUMBER
　　　　ASSIGNMENT 114
　　6.9.1   Engineering Procedure Number Assignment Log ... 114

6.9.2   Example—Engineering Procedure Number
                Assignment Log 116
6.10.0 ENGINEERING PROCEDURE TITLES 116
6.11.0 FORM NUMBERS 118
        6.11.1 Example—Form Numbers and Interpretation 118
6.12.0 FORM NUMBER ASSIGNMENT 119
        6.12.1 Example—Form Number Assignment Log 119
6.13.0 MANUAL NUMBERS 122
        6.13.1 Manual Numbers and Interpretation 123
6.14.0 MANUAL NUMBER ASSIGNMENT 123
        6.14.1 Example—Manual Number Assignment Log 123
6.15.0 DOCUMENT REVIEW FORM NUMBERS 126
        6.15.1 Example—Document Review Form Numbers
                and Interpretation 127
6.16.0 DOCUMENT REVIEW FORM NUMBER
        ASSIGNMENT 127
        6.16.1 Document Review Form Number
                Assignment Log 128
6.17.0 DOCUMENT CHANGE REQUEST FORM NUMBERS 131
        6.17.1 Example—Document Change Request Form
                Numbers and Interpretation 132
6.18.0 DOCUMENT CHANGE REQUEST FORM
        NUMBER ASSIGNMENT 132
        6.18.1 Document Change Request Form Number
                Assignment Log 132
        6.18.2 Example—Change Request Form Number
                Assignment Log 134
6.19.0 DOCUMENT MASTER LIST 134
        6.19.1 Example—Document Master List 136

7   **Electronic Database** **139**
7.1.0   DOCUMENT NUMBER STRUCTURE 139
        7.1.1   Prefix 140
        7.1.2   Suffix 140
        7.1.3   Identification Numbers and Interpretation 140
7.2.0   DOCUMENT DATABASE 141
        7.2.1   Example—Document Database 142

Appendix A: Policy Example **145**

Appendix B: Departmental Instructions Example **148**

Appendix C: Engineering Procedure Example **156**

Appendix D: Forms, Lists, and Logs **163**
    Controlled Document Agreement 164
    Engineering Procedure Manual Distribution List 165
    Document Review Form 166
    Document Change Request Form 166
    Policy Number Assignment Log 168
    Departmental Instruction Number Assignment Log 169
    Engineering Procedure Number Assignment Log 170
    Form Number Assignment Log 171
    Manual Number Assignment Log 172
    Document Review Form Number Assignment Log 173
    Document Change Request Form No. Assignment Log 174
    Document Master List 175
    Document Database 176

Appendix E: Further Reading **177**
    WORLD WIDE WEB REFERENCES 178

Index **179**

1

Introduction

Where does it say that you have to develop, change, review, approve, and follow Engineering Procedures?

Well, if you are or want to be in compliance with ISO 9000 regulations, then the following sections apply to you.

ISO 9001 Sec. 4.4, Design Control:

> The supplier shall establish and maintain procedures to control and verify the design of the product in order to assure that the specified requirements are met.

ISO 9001 Sec. 4.5, Document Control; Subsection 4.5.1, Document Approval and Issue:

> The supplier shall establish and maintain procedures to control all documents and data that relate to the requirements of this international standard. These documents shall be reviewed and approved for adequacy by authorized personal prior to issue. This control shall ensure that:
>
> a) The pertinent issues of appropriate documents are available at all locations where operations essential to the effective of the quality system are performed.
>
> b) Obsolete documents are promptly removed from all points of issue or use.

ISO 9001 Sec. 4.5, Document Control; Subsection 4.5.2, Document Changes/Modifications:

> Changes to documents shall be reviewed and approved by the same functions/organizations that performed the original review and approval unless specifically designated otherwise. The designated organizations shall have access to pertinent background information upon which to base their review and approval.
>
> Where practicable, the nature of the change shall be identified in the document or appropriate attachments.
>
> A master list or equivalent document control procedure shall be established to identify the current revision of documents in order to preclude the use of non-applicable documents.

Documents shall be re-issued after a practical number of changes have been made.

Besides ISO 9000 there are several other reasons why you will need to document your engineering operation:

- Customers will impose their own documentation requirements on you.

- Sometimes you will need to work to Military Standards.

- The Food and Drug Administration have their *current good manufacturing practices* (cGMP).

- If you are going to do business with international suppliers, ISO 9000 is the way to go.

- Or it is just a good idea?

Regardless of the reason for which you are documenting your operation, the unexpected benefits will be greater operational efficiency, increased profitability and savings in administrative costs, and improvements in marketing and sales activity. Putting together an engineering documentation system is a great aid in determining exactly how you are operating now, where you need to improve, and the best way to carry out your processes.

This guide book consists of twenty-five engineering procedures and five forms that can be used by any company to establish an engineering procedural documentation system. The first part of this book covers a new company setting up an engineering procedures (paper based) documentation system and Ch. 7 covers an electronic database. The paper based system uses engineering procedure manuals as a method of distributing new and revised procedures to end-users. All of the methods shown in this book can be tailored to fit each company's unique operation. This book is written in such a way that those employees responsible for writing and managing engineering procedures can use it as a guideline to perform their daily tasks. The table of contents follows the normal sequence of an engineering procedure documentation process and documentation methods are established in each of the chapters. The document formats and numbering systems can be used by most companies, and they are acceptable per ISO 9000 guidelines. Several of the engineering procedures include forms that will aid in the information gathering process.

1.1.0 ENGINEERING PROCEDURES CASE STUDY

Once upon a time there was a manufacturing company that was manufacturing products long before ISO 9000 regulations ever existed. They were documenting some of their procedural methods but they still did not have complete confidence that their process was as good as it could be. When ISO 9000 regulations came into being, they stepped up their engineering procedures writing activities to be compliant with the guidelines.

The existing procedures were analyzed for quantity, and evaluated for quality and availability. After performing an analysis of the document system, the consensus was that there were not enough procedures and they were all different because there was not a standard format in place to follow.

Some of the documents were not even engineering procedures. As you can imagine, the people who were tasked with writing engineering procedures had a full time job. They could only write them in their spare time. As far as control goes, it would take from two to six months to get an engineering procedure through the system. After some time had passed, their documentation system received a boost when they organized a continuous improvement team to reengineer their engineering procedures system. An analysis and summary were prepared and, as a result, corrective action plans were developed and implemented. The document system was modernized and within six months the company had an easy-to-use and easy-to-maintain documentation system. This is where this book comes in, it covers everything it takes to develop and manage engineering procedures and their associated documentation.

1.2.0   ENGINEERING PROCEDURES MANAGEMENT

This book shows how to develop and manage engineering procedures and their associated documents for all engineering department functions (Fig. 1.1). The engineering department is responsible for generating and maintaining procedures governing its day-to-day operations. The released procedures will be contained in manuals for ease of use and revision. Each manager will approve new and revised procedures for his/her particular function. Managers of other departments affected by the procedure must also approve the documents. The engineering department is also responsible for the producing, distributing, and maintaining procedure manuals. This book is written in the order that documentation functions are carried out.

✓ First, establish a procedures writing group, then identify which documents need to be developed.

✓ Second, establish a method for writing each type of document and develop formats and contents.

✓ Third, after documents are approved they will be released into engineering document control. Engineering document control will produce the engineering procedure manuals and distribute them to end-users.

✓ Fourth, a change control system will need to be developed to maintain the integrity of the system.

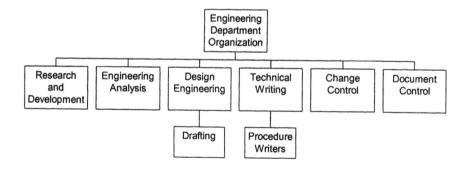

Figure 1.1. Engineering department organization.

1.3.0   DOCUMENT CONTROL SYSTEM

The document control system (Fig. 1.2) is the main focus of any procedural effort, and it controls the framework of the entire process. Engineering procedures must be developed and approved, prior to use, per ISO 9000 requirements. The challenge today is to develop and control documentation in the most effective and efficient way possible. This book provides steps in that direction.

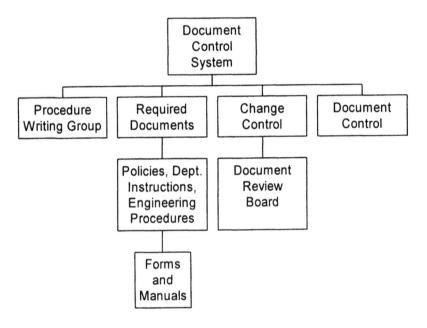

Figure 1.2. Engineering document control system.

1.4.0   DOCUMENT CONTROL COMPONENTS

The following list assumes that a company is starting an engineering procedure documentation effort from scratch. If a documentation system already exists, compare this list with your own to see if all the bases have been covered. All of the documentation components have been listed to show a complete documentation system.

— Establish a procedure writing group
— Set up an engineering procedure documentation system
— Identify which documents to write first, then most important to less important
— Establish document format and contents
— Develop forms
— Develop engineering procedure manuals
— Assign writing tasks to documentation writers
— Write policies, departmental instructions, and engineering procedures
— Assign numbers and names
— Establish a reviewers and approvers
— Establish a change control system
— Establish a Document Review Board
— Assign change control numbers
— Establish a document control system
— Establish a distribution system

1.5.0 WHAT EXACTLY DO YOU NEED TO DOCUMENT?

The document control requirements are fully described in ISO 9001 Standard. The traditional four-tier documentation pyramid for ISO 9000 is:

> 1st tier...Quality Manual
>
> 2nd tier...Procedures
>
> 3rd tier...Work Instructions
>
> 4th tier... Forms and Records

Following is my take on the traditional documentation system. My system begins with policies, departmental instructions, procedures and forms, and when the forms are filled in, they become records. Figure 1.3 illustrates the key documents that will need to be developed and retained to support the previously listed documentation efforts. This documentation is required to operate as an engineering department. See Ch. 3 and Appendices A, B, and C for examples of each document type.

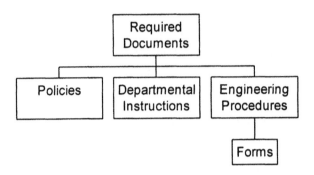

Figure 1.3. Required documents.

1.6.0   ENGINEERING PROCEDURE WRITING

The simplest method of organizing the engineering procedure documentation effort is to have engineering personnel write, change, and control their own operating procedures, but in most companies, operating procedures are written by a centralized group of procedure writers. This group writes the operating procedures for all of the departments in the company.

The functions of the centralized procedure writing group are:

For new engineering procedures:

Interview engineering department employees

Write new procedures

Assign numbers

Obtain Manager approval

Send out for review and approval

Hold Document Review Board meetings

Release approved documents to engineering document control

For changes to engineering procedures:

Receive requests for changes

Prepare document change package

Obtain Manager approval

Send out for review and approval

Incorporate changes

Hold Document Review Board meetings

Release changed documents to engineering document control

1.7.0 HOW TO USE THIS BOOK

This book covers the development and management of engineering procedures used to define the engineering department functions. This book can be used in several ways, for example, when preparing to write a policy, departmental instruction, or engineering procedure, you can look in Ch. 3 and Appendices A, B, and C.

Next, you can tailor these documents to fit your unique procedural document situation. Because the templates are already filled, you are given concrete examples of how to write the documentation. If you have a subject in mind, such as forms, you can go to the table of contents, or, for a more detailed explanation of the subject, use the index. This book can be used for developing and managing engineering procedure documentation and for training new employees or retraining existing employees.

2

Engineering Procedure
Writing Group

This chapter covers the method of setting up a new engineering procedure writing group and the steps that all procedures go through, and how to write procedures. New companies starting an engineering procedure writing effort from scratch will need to include procedure writers in the technical writing group (see Fig. 2.1). This book covers the method where engineering procedure writers write their own policies, departmental instructions and engineering procedures and obtain approval from the Document Review Board (Ch. 5) and releases them into engineering document control. Engineering document control publishes the manuals and distributes them to end-users. The engineering procedure writers perform all of the subsequent changes to the policies, departmental instructions and engineering procedures and distribute changed pages to end-users to be inserted into their manuals.

In larger companies, the procedure writers report to corporate communications. With this method, a department outside of engineering writes all of the procedures and releases them into document control where they are copied and distributed.

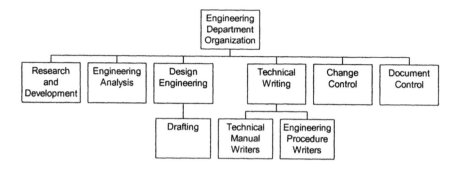

Figure 2.1. Engineering procedure writers.

First, you will need to determine your reason for creating an engineering procedure writing group. Following are some examples:

- You need to document the engineering department's operating methods

- You need to re-engineer your existing engineering documentation system

Why document the engineering documentation system in the first place? Here are some good reasons why:

- Customer contracts

- Government regulations

- It's a good idea

2.1.0   HOW TO ESTABLISH AN ENGINEERING PROCEDURE WRITING GROUP

Send out a memo (Fig. 2.2) that explains the function of the procedure writing group.

MEMO

To:

From:

Subject: Engineering Procedures Writing Group Formation

In a endeavor to have every engineering department employee follow a standard common reference system, and to give direction to employees concerning their unique engineering function, policies, departmental instructions, and engineering procedures are going to be developed and implemented. Technical writing has been assigned the responsibility of organizing and implementing the following publications:

Policy Manual

Departmental Instruction Manual

Engineering Procedures Manual

The manuals will contain information necessary to provide uniformity, standardize definitions, and clarify department responsibilities. The manuals will designate forms or documents to be utilized when necessary and state procedural steps to be followed to assure consistency of action and effect overall coordination.

Manual sections will be released as they are approved eventually resulting in complete manual that will be used throughout the company. To release sections of these manuals, individuals who have expertise in certain areas will be asked to provide input.

Distribution:

Figure 2.2. Engineering procedure writing group memo.

2.1.1   Kick-off Meeting Memo

MEMO

To:

From:

Subject: Kick-off Meeting

There will be a kick-off meeting for the formation of a new Engineering Procedures Writing Group and the Document Review Board that will review and approve all documentation.

We will be discussing how the Document Review Board will operate and answer any questions you may have regarding this new Board.

Distribution:

Figure 2.3. Kick-off meeting memo.

2.1.2   Agenda for the Kick-off Meeting

◆   Purpose of the Engineering Procedures Writing Group

    Review all new or revised policies, departmental instructions, and engineering procedures.

    Resolve document discrepancies.

    Authorize documents for release.

◆   Individual Reviewer Responsibilities

    Attend Document Review Board meetings when scheduled.

    Provide input on documents.

    Coordinate review with knowledgeable persons.

Other individuals who have expertise in their respective areas will be asked to attend meetings and provide input when necessary.

◆ Procedure for Document Release

> A draft of each document will be sent out for review
>
> A one or two week review time period
>
> Reviewers meet to discuss document mark-ups
>
> Reviewers resolve any discrepancies
>
> Reviewers sign Document Review form
>
> Released documents are distributed to all manual holders

◆ Schedule of Document Review Board Meetings

> Meet every week for one hour
>
> Time and date to be announced in previous meeting for the next meeting

2.2.0 BASIC WRITING GROUP FUNCTIONS

2.2.1 Which Documents are Necessary?

After establishing the writing group, the first order of business is to determine which documents need to be written and maintained. If this is a new company you could hold several meetings where you establish lists of the documents you think you might need. Next, you will have to prioritize the documents to make sure that the most important documents are identified and worked on first. Having a lower priority does not mean that the documents are not needed; they are just not associated directly with the end product. If this is an existing company, you may need to re-engineer your documentation system. First, review your existing documents and develop a list of which ones you think you might need to develop, revise or obsolete to make your documentation system more useful. Following are some questions that the procedure writing group will need to answer:

Who will write new procedures?

Who can request that a new procedure be written?

Who determines if the request for a new procedure is valid?

Who edits procedures?

Who reviews procedures?

Who approves procedures?

Who are the members of the Document Review Board?

Who can request changes to procedures?

Who incorporates change requests?

Who releases procedures?

Who keeps records of procedures?

Who produces procedure manuals?

Who distributes manuals to end-users?

Who updates the manuals?

2.2.2   Establish Document Format and Contents

Next, establish the format and contents for each document type. You need to develop a document format standard to make each document uniform. For a new company, you will not have any examples to follow except one that you might not realize that you have. You might have a house style. House style means that someone (usually in Administration or Marketing) has established the company's image such as logo, stationary, forms, memos, etc., therefore, you will have to stay within those guidelines. There are several ways to obtain examples of your new documents:

- I always bring examples from other companies where I have worked.

- I send out a memo to the entire company asking if anyone has examples of policies, departmental instructions or engineering procedures for other companies where they have worked.

- I contact companies in the area and ask for examples. This actually works.

- Or you can use Ch. 3 and Appendices A, B, and C of this book where I have examples of each of the document's format and contents.

- Also see Appendix E where I have listed several books and World Wide Web sites that show examples of all of the documents mentioned above.

2.3.0   BEFORE YOU START WRITING

Now that you have established your format and contents, you can start collecting the information to write the documents that you have identified. You can obtain the information through interviewing, or you might know the subject well enough to write the document yourself.

2.3.1   Interviewing

By now everybody has heard about the new documentation effort that you are involved in and there should not be any surprises. All you have to do is find the person that has the information that you need to write your document and schedule an interview time. Have a list of questions to follow so that you do not miss any of the information that you might need to write your document. Other information will come from procedures that were written by the employee or their manager that they have been using for years. After you have written a draft of the document, send a copy to the person that you interviewed for an edit. This will get them involved in the process. Following are some example questions that you can ask at the interview.

2.3.2   Questions to ask for a New Procedure

Who uses the procedure?

What is the procedure for?

When is the procedure used?

Where is the procedure used?

Why is the procedure needed?

How is the procedure used?

2.3.3   Questions to ask for a New Form

Who uses the form?

What is the form for?

When is the form used?

Where is the form used?

Why is the form needed?

How is the form used?

2.4.0   POLICY WRITING TIPS

Why would you want to write a policy in the first place? Let's say you are going to start a new company. Would you start off your documentation system by writing policies? I don't think so. You would write the documents that will help you make money, now.

> ➢ The first documents that you would write would be the ones that are used to manufacture your product. Most of these documents will be used across several departments and will require review and approval from them.

> ➢ After some time has passed you would start writing departmental instructions to define how employees perform their departmental functions. These documents are reviewed and approved by the immediate department.

> ➢ Then after several successful years of operation you might want to start writing policies that will convey the intent of the engineering department to operate in a generalized way. Example: there will be an Engineering Department that develops new products and changes existing products.

Who are policies written for? Policies are written as a general record of the engineering department's common purpose or intent. Another way to say the same thing would be "The policy is used to provide a statement of intent."

According to the American Heritage Dictionary, a *policy* is: *1) a plan or course of action, as of a business, intended to influence and determine decisions, actions, and other matters; 2) a course of action, guiding principle, or procedure considered expedient, prudent, or advantageous.*

Engineering policies are executive management's commitment to and involvement in engineering functions. Policies define the basic values of the company for all of the decisions involving engineering and the company. Through the establishment of and adherence to policies, management obtains employee commitment to engineering department functions. Policies are essential documents that are put in place to assist management to be consistent in the way it conducts business. Engineering procedures naturally flow from the policies. Once a policy has been defined, the associated engineering procedure will ensure the proper execution of the policy. Engineering policies provide employees at all levels with the information to consistently carry out their responsibilities.

2.4.1   How Many Policies are Really Needed?

How do you decide how many policies you will need? Your guess is as good as mine, but generally there should be at lease one policy that covers the entire engineering operation. Then, there will need to be a few system overview policies that convey the engineering department's plan of operation, and lastly, some of the subfunctions could be documented.

There could be policies for the following engineering systems:

- Product Development
- Product Design
- Product Phases
- Product and Document Identification
- Required Documents
- Customer Documents
- Vendor Documents
- Document Change Control
- Document Control

Next, there could be several subfunctions such as:

- Research and Development (R&D)
- Engineering Analysis
- Design Engineering
- Drafting
- Technical Writing
- Change Control
- Document Control

Here is an example of some of the items that might be in a policy for the Research and Development portion of an engineering department.

We intend to have an engineering department that will perform research and development activities to development new products for introduction into the existing product stream. The Research and Development department will perform some of the following actions to accomplish this endeavor:

Perform new product planning

Hold Product Review Board meetings

Meet with the Make-or-Buy Committee to decide the best method of manufacturing

Address safety requirements

Develop preliminary specifications, drawings, and bills of materials

Perform stress analysis

Hold design review meetings

Purchase or make parts and assemblies

Develop a prototype of the product

Perform product testing

Prepare final specifications, drawings, and bills of materials

Release design into manufacturing for production

Perform liaison with manufacturing

2.5.0 DEPARTMENTAL INSTRUCTION WRITING TIPS

Departmental Instructions (also known as administrative procedures) are used to document administrative duties within the engineering department. They only require the originators' and manager's signatures and change control is informal. The goal is to have every employee follow the same procedure when performing routine tasks. Departmental instructions are usually routine tasks, such as, computer input, document control filing, etc. Departmental Instructions contain information necessary to provide engineering uniformity, standardize definitions, and clarify engineering responsibilities. They also designate forms or documents to be utilized when necessary and state procedural steps to be followed to assure consistency of action and create overall coordination.

According to the American Heritage Dictionary, a *departmental instruction* is: *1) the act, practice, or profession of instructing; 2a) imparted knowledge, 2b) an imparted or acquired item of knowledge, a lesson.*

2.5.1 How Many Departmental Instructions are Really Needed?

How do you decide how many departmental instructions are required? Generally, there should be one for each department activity that needs to be standardized. Some examples would be:

- How to load documentation information into the database
- How to assign the next number to a document

2.6.0 ENGINEERING PROCEDURE WRITING TIPS

This section covers the format and contents of engineering procedures. The ISO 9000 quality documentation standard states that operations must be documented and maintained. Engineering procedures will need to be written for all operations or systems if you are working towards or have been ISO 9000 certified.

2.6.1   Definitions

According to the American Heritage Dictionary, an *engineering procedure* is: *1) a manner of proceeding, a way of performing or effecting something; 2) a series of steps taken to accomplish an end.* Other definitions include:

- Authorized and controlled step-by-step instructions that describe how to perform tasks to reach a specified goal.

- Written procedures prescribing and describing the steps to be taken in normal and defined conditions, which are necessary to assure control of production and processes.

Engineering procedures are the step-by-step instructions that employees follow to fulfill the requirements of each job. They are the steps that must be followed for a job to be done correctly.

If the company does not have any engineering procedures and you are going to suggest that they need them, the first question that upper management will ask is, "Why do we need engineering procedures now? We have gotten along all these years without them!" Here are some good reasons why:

- Engineering procedures give the end users a standard frame of reference.

- The methods of operation need to be documented so that the information is not just stored in someone's head. This way the information is not lost when people leave the company or change jobs.

- Engineering procedures housed in manuals are easy to access and you can find answers to questions faster.

- In some cases engineering procedures are required to comply with regulatory agencies guidelines and industry standards such as ISO 9000.

- Having engineering procedures documented saves time and guarantees accurate responses.

- Engineering procedures can be used as training tools for new employees and retraining existing employees.

How do you decide if you need an engineering procedure? The absence of a procedure would have an adverse affect on safety, cost, or schedule.

3

Document Format and Contents

3.1.0 POLICY FORMAT

Figure 3.1 illustrates the format of a policy. The policy has the simplest format compared to procedures. The cover page could have a header and a footer to capture useful information and approval signatures and subsequent pages will just have a header to link the pages together. The same format is used for all policies, because having them all the same, as much as possible, is more economical. Any time there are variations in thinking patterns, time and money are wasted.

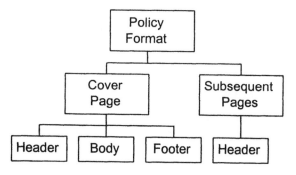

Figure 3.1. Policy format.

This section shows how to develop a complete policy. The policy is used to convey the date that it was released for use, and who prepared and approved it. With a policy you will be answering the following questions:

- What is the policy title?
- What is the policy number?
- What date was the policy released?
- How many pages are there?
- Who prepared the policy?
- Who reviewed and approved the policy?

Policy (Page 1). The following is an example of the first page of a policy. Identification information will need to be recorded on the first page, such as the policy number, page number, and the signature of the originator and approvers.

Policy (Page 1) - Preparation. Each of the following circled numbers corresponds to the circled numbers on page 1 of the example policy.

❶ Title: Enter the title.
   Example: Make or Buy Policy

❷ Policy No.: Enter the policy number. (See Ch. 6 under Policy Numbers.)
   Example: P-05

❸ Release Date: Enter the date that the policy was released for use.

❹ Page: Enter the page number.
   Example: Page 1 of 3. This method is best for change control purposes.

❺ Prepared By: Enter your name and the date when the policy is complete.

❻ Approved By: Enter your name and the date when your review and approval are complete.

Policy (Page 2). Following is an example of the second page of a policy. All of the information in the header is the same as page one. Signatures are not required on subsequent pages; therefore, a footer is not required.

⊕	Policy	
Title: ❶		Policy No.: ❷
Release Date: ❸		Page: ❹
	Cover Page	
❺ Prepared By:		
Engineer:		Date:
❻ Approved By:		
Engineering Manager:		Date:
Vice President Engineering:		Date:

Example–Policy Page 1

🌎	Policy	
Title: ❶		Policy No.: ❷
Release Date: ❸		Page: ❹

Subsequent Pages

Example–Policy Page 2

Policy (Page 2) - Preparation. Each of the following circled numbers below corresponds to the circled numbers on the subsequent pages of the example policy.

❶ Title: Enter the title. Same as page 1.
Example: Make or Buy Policy

❷ Policy No.: Enter the policy number. Same as page 1.
Example: P-01

❸ Release Date: Enter the date that the policy was released for use. Same as page 1.

❹ Page: Enter the page number.
Example: Page 2 of 3. Specify the number of the page sequence.

3.2.0   POLICY CONTENT

Figure 3.2 illustrates the required sections of a policy. Policies must be written in a format that is clear and easy to understand so that nothing is left to interpretation. Use the company house style if one exists. If the company does not have a style, the policy examples in this book can be used as ISO 9000 guidelines, and they should be tailored to meet individual company requirements.

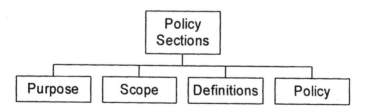

Figure 3.2. Policy sections.

3.3.0   EXAMPLE POLICY

See Appendix A for another example of a policy.

(🌐)	**Policy**	
Title: Make or Buy Policy		Policy No.: P-01
Release Date: 10/25/00		Page: 1 of 2

1.0   Purpose

The purpose of this policy is to provide guidance for engineering personnel in making the decision of whether to make or buy a part or an assembly.

2.0   Scope

This policy applies to all personnel involved in make-or-buy decisions.

3.0   Definitions

Make Item: An item produced or work performed by the company. Buy Item: An item that is produced or work performed outside the company.

Must Make Item: An item or service which the company regularly makes and is not available (quality, quantity, delivery, and other essential factors considered) from outside vendors. Must Buy Item: An item or service that the company does not have the capacity to provide.

Can Make-or-Buy Item: An item or service that can be provided by the company or outside vendor at prices comparable to the company (considering quality, quantity, delivery, risk, and other essential factors.)

Prepared By:	
Engineer:	Date:

Approved By:	
Engineering Manager:	Date:
Vice President Engineering:	Date:

Example–Policy Page 1

🌐	**Policy**	
Title: Make or Buy Policy		Policy No.: P-01
Release Date: 10/25/00		Page: 2 of 2

4.0 Policy

4.1 Make-or-Buy Decisions

Make-or-buy decisions are to be based on good business practices. The following factors are considered in arriving at make-or-buy decisions.

Total cost

Schedule considerations, including risk

Quality of product or service

Performance of item or service

Complexity of item or difficulty of administering control

Loading of functional organizations which would normally perform the tasks to assure that adequate capacity exists

Facilities and capital equipment requirements

Availability of competition, especially from small business firms and labor surplus area firms

Quality of technical data package used for procurement of an item or service

Technical and financial risks associated with potential suppliers

4.2 Make-or-Buy Committee

There will be a Make-or-Buy Committee that will evaluate and decide on all Make-or-Buy plans, items and services presented to it.

The Committee consists of the following members or their designees.

Project Manager – Chairperson

Engineering Manager

Manufacturing Manager

Quality Manager

Purchasing Manager

Finance Manager

Contract Administration

Example–Policy Page 2

3.4.0   DEPARTMENTAL INSTRUCTION FORMAT

Figure 3.3 illustrates the format of a departmental instruction. Departmental Instructions have a simpler format than engineering procedures. The cover page can have a header and a footer to capture useful information and approval signatures and subsequent pages will just have a header to link the pages together. The same format is used for all departmental instructions, because having them all the same, as much as possible, is more economical. Any time there are variations in thinking patterns, time and money are wasted.

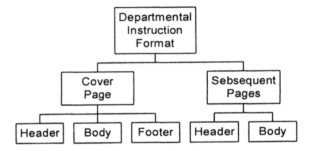

Figure 3.3. Departmental instruction format.

This section shows how to develop a complete departmental instruction. A departmental instruction is used to convey the date that it was released for use, and who prepared and approved it. With departmental instructions you will be answering the following questions:

- What is the title?
- What is the identification number?
- What date was it released?
- How many pages are there?
- Who prepared the departmental instruction?
- Who reviewed and approved the departmental instruction?

Departmental Instruction (Page 1). Following is an example of the first page of a departmental instruction. Identification information will need to be recorded on the first page, such as the departmental instruction number, page number, and the signatures of the originator and approver.

(🌍) Departmental Instruction		
Title: ❶		Dept. Inst. No.: ❷
Release Date: ❸		Page: ❹
	Cover Page	
❺ **Prepared By:**		
Engineer:		Date:
❻ **Approved By:**		
Engineering Manager:		Date:

Example–Departmental Instruction Page 1

Departmental Instruction (Page 1) - Preparation. Each of the following circled numbers corresponds to the circled numbers on page one of the example departmental instruction.

❶ Title: Enter the title.

Example: Master File Update

❷ Dept. Inst. No.: Enter the departmental instruction number. (See Ch. 6 under Departmental Instruction Numbers.)

Example: D-01

❸ Release Date: Enter the date that the departmental instruction was released for use.

❹ Page: Enter the page number.

Example: Page 1 of 3. This method is best for change control purposes.

❺ Prepared By: Enter your name and the date when the departmental instruction is complete.

❻ Approved By: Enter your name and the date when your review and approval is complete.

Departmental Instruction (Page 2). Following is an example of the second page of a departmental instruction. All of the information in the header is the same as page one. Signatures are not required on subsequent pages; therefore, a footer is not required.

(🌐)	**Departmental Instruction**	
Title: ❶		Dept. Inst. No.: ❷
Release Date: ❸		Page: ❹

Subsequent Pages

Example–Departmental Instruction Page 2

Departmental Instruction (Page 2) - Preparation. Each of the following circled numbers below corresponds to the circled numbers on the subsequent pages of the example departmental instruction.

❶ Title: Enter the title. Same as page 1.

Example: Master File Update

❷ Dept. Inst. No.: Enter the departmental instruction number. Same as page 1.

Example: D-01

❸ Release Date: Enter the date that the departmental instruction was released for use. Same as page 1.

❹ Page: Enter the page number.

Example: Page 2 of 3. Specify the number of the page sequence.

3.5.0   DEPARTMENTAL INSTRUCTION CONTENTS

Figure 3.4 illustrates the required sections of a departmental instruction. Departmental instructions must be written in a format that is clear and easy to understand where nothing is left to interpretation. Use the company house style if one exists. If the company does not have a style, the departmental instruction examples in this book can be used as ISO 9000 guidelines, and they should be tailored to meet individual company requirements.

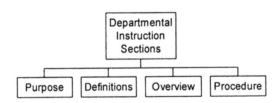

Figure 3.4. Departmental instruction sections.

3.6.0   EXAMPLE DEPARTMENTAL INSTRUCTION

See Appendix B for another example of a departmental instruction.

⊕	**Departmental Instruction**
Title: Where-Used Input	Dept. Inst. No.: D-01
Release Date: 6/19/00	Page: 1 of 5

1.0   Purpose

The purpose of this procedure is to provide instructions, and assign responsibilities for inputting the Where-Used information, for parts and assemblies, into the computer.

2.0   Definitions

Class 1 Change-Interchangeable: A part (1) possesses such physical characteristics as to be equivalent in reliability, and maintainability, to another part; and (2) is capable of being exchanged for the other item (a) without alteration for fit, and (b) without alteration of adjoining items.

Class 2 Change-Non-Interchangeable: A change that is not interchangeable, that affects form, fit and function.

Form – The manufactured structure, shape, and material composition of an item or assembly.

Fit – The size and dimensional aspect of an item or assembly.

Function – The actual performance and level of performance of an item or assembly.

Class 3 Change-Documentation Only: Documentation is used to indicate changes to documentation, drawings, and bills of materials, etc., which are strictly error corrections and do not affect the manufacturing process.

Prepared By:	
Engineer:	Date:
Approved By:	
Engineering Manager:	Date:

Example–Departmental Instruction Page 1

🌀	**Departmental Instruction**	
Title: Where-Used Input		Dept. Inst. No.: D-01
Release Date: 6/19/00		Page: 2 of 5

3.0   Overview

3.1   A Where-Used check will be performed on all Class 1 or 2 changes, and Class 3 changes which are noted as "Must Trace." Engineering will be responsible for the accuracy of the Where-Used research and will perform the Where-Used look-ups when:

> Processing an Engineering Change that changes an old part number to inactive.

> Processing Class 1 or 2 Engineering Changes.

All part numbers that are changed must have a Where-Used performed.

3.2   The originator of an Engineering Change is responsible for obtaining a Where-Used look-up prior to starting a Change Control Board meeting. The Where-Used must be present at the Change Control Board meeting.

3.3   The Engineering Change Cover Sheet will be verified to insure that the products listed on the cover sheet are consistent with the results of the Where-Used in determining the Top Levels (product types) that are affected by the change.

3.4   All old numbers noted on the Where-Used reflect the product, model, feature/option and customer name of the next assemblies that may be affected by the change.

3.5   If all applications of the old part numbers are to be affected by the change, then the Engineering Change Cover Sheet should be marked accordingly.

3.6   If all of the applications of the old part numbers that are to be affected by the change are not impacted, then the Engineering Change Cover Sheet will be marked accordingly. Justification for excluding certain part numbers from the change must be noted in the Where-Used by the originator of the Engineering Change.

3.7   The Bill of Material is the report that will be the source of most information on the Where-Used. Customer names will be obtained from the Engineering Change.

Example–Departmental Instruction Page2

⊛	**Departmental Instruction**
Title: Where-Used Input	Dept. Inst. No.: D-01
Release Date: 6/19/00	Page: 3 of 5

4.0   Where-Used Report Input Procedure

This procedure is divided into the following parts:

   • Where-Used Input Procedure

   • How to Prepare the Where-Used Input Form

The following procedure describes who is responsible and what they are supposed to do for each processing step.

4.1   Engineering

Steps

   1. Prepare the Where-Used Input form when there is an Engineering Change that might affect the location of parts or assemblies within the product structure.

   2. Copy, distribute and file the completed Where-Used Input form.

Example–Departmental Instruction Page 3

⊕	**Departmental Instruction**
Title: Where-Used Input	Dept. Inst. No.: D-01
Release Date: 6/19/00	Page: 4 of 5

5.0   Overview

The procedure for processing the Where-Used Input form shall be followed by each individual responsible for entering information on the form. Each circled number below corresponds to the circled number on the following Where-Used Input form E026.

5.1   Engineer

❶ Enter your name and date that the Where-Used Input Form was completed.

❷ Enter the associated Engineering Change number.

5.2   Engineering Manager

❸ Enter your name and date upon approval of the infomation entered on the form.

5.3   Engineer

❹ Enter: The part number of the material currently being used that will be replaced, reworked, or scrapped by the proposed change.

A brief description of the old part number.

The assembly number(s) into which the old part number is built. The first level assembly into which the old part goes.

A brief description of the next assembly part number.

Example–Departmental Instruction Page 4

(🌐)	**Departmental Instruction**	
Title: Where-Used Input		Dept. Inst. No.: D-01
Release Date: 6/19/00		Page: 5 of 5

5.3   **Engineer** (Cont'd.)

Enter: (Cont'd.)

The first Top Level that appears for the next assembly(s). The top level part number is noted in the product structure tree (drawing) or configuration control document, basic units, features and options.

A brief description of the top-level part number that identifies the unique feature/option.

The model number within the product family that identifies the product types.

The product family number that defines which family the model type is from.

Example–Departmental Instruction Page 5

Where-Used Input		
Engineer: ❶	Date:	EC No.: ❷
Engineering Manager: ❸		Date:
Old Part Number	❹	
Description		
Quantity		
Next Assembly		
Description		
Top Level Part Number		
Feature/Option		
Model Number		
Product		
Customer		

Form E026 (Procedure EP-2-10) 5/28/00

Example–Where-Used Input

3.7.0   ENGINEERING PROCEDURE FORMAT

Figure 3.5 illustrates the format of an engineering procedure. The cover page can have a header and a footer to capture useful information and approval signatures and subsequent pages will just have a header to link the pages together. The same format is used for all engineering procedures, because having them all the same, as much as possible, is more economical. Any time there are variations in thinking patterns, time and money are wasted.

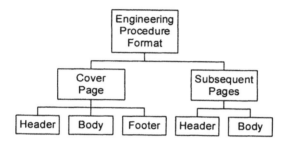

Figure 3.5. Engineering procedure format.

This section shows how to develop a complete engineering procedure. The engineering procedure is used to convey the date that it was released for use, and who prepared and approved it. With an engineering procedure you will be answering the following questions:

- What is the title?
- What is the identification number?
- What date was it released?
- How many pages are there?
- Who prepared the engineering procedure?
- Who reviewed and approved the engineering procedure?

Engineering Procedure (Page 1). Following is an example of the first page of an engineering procedure. Identification information will need to be recorded on the first page, such as the engineering procedure number, page number, and the signatures of the originator and approver.

Engineering Procedure			
Title: ❶		Eng. Proc. No.: ❷	
Release Date: ❸	Revision: ❹		Page: ❺

Cover Page

❻ Prepared By:	
Engineer:	Date:
❼ Approved By:	
Engineering Manager:	Date:
Manufacturing Manager:	Date:
Quality Manager:	Date:

Example–Engineering Procedure Page 1

Engineering Procedure (Page 1) - Preparation. Each of the following circled numbers corresponds to the circled number on page 1 of the example engineering procedure.

❶ Title: Enter the procedure title.

Example: Engineering Change Proposal

❷ Eng. Proc. No.: Enter the procedure number. (See Ch. 6 under Engineering Procedure Numbers.)

Example: EP-04-02

❸ Release Date: Enter the date that the engineering procedure was released for use.

❹ Revision: Enter the revision letter. "A" will be used for the first revision, and "B" for the next, and so on.

❺ Page: Enter the page number.

Example: Page 1 of 9.

This method is best for change control purposes.

❻ Prepared By: Enter your name and the date when the engineering procedure is complete.

❼ Approved By: Enter your name and the date when your review and approval is complete.

Engineering Procedure (Page 2). Following is an example of the second page of an engineering procedure. All of the information in the header is the same as page one. Signatures are not required on subsequent pages; therefore, a footer is not required.

	Engineering Procedure	
Title: ❶		Eng. Proc. No.: ❷
Release Date: ❸	Revision: ❹	Page: ❺

Subsequent Page

Example–Engineering Procedure Page 2

Engineering Procedure (Page 2) - Preparation. Each of the following circled numbers corresponds to the circled numbers on the subsequent pages of the example engineering procedure.

❶ Title: Enter the procedure title. Same as page 1.
Example: Engineering Change Proposal

❷ Eng. Proc. No.: Enter the engineering procedure number. Same as page 1.
Example: EP-04-02

❸ Release Date: Enter the date that the engineering procedure was released for use. Same as page 1.

❹ Revision: Enter the revision letter. The revision letter could be different than page 1.

❺ Page: Enter the page number.
Example: Page 5 of 9. Specify the number of the page in the sequence.

3.8.0   ENGINEERING PROCEDURE CONTENTS

Figure 3.6 illustrates the required contents of an engineering procedure. The engineering procedure examples in this book can be used as ISO 9000 compliance guidelines, and they should be tailored to meet individual company requirements. The following pages show an example of a complete engineering procedure.

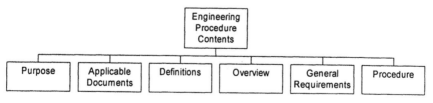

Figure 3.6. Engineering procedure contents.

Not all of the sections are required in each engineering procedure. The three sections: Purpose, Applicable Documents, and Procedure, are mandatory and the other three: Definitions, Overview, and General Requirements, can be used as necessary. If any of the mandatory sections are not used, enter the section heading and the words "Not used" under that heading.

Purpose: The purpose describes in general terms the intent of the procedure. List the number and title of each procedure and form. A complete exhibit of each form shall be included in the procedure. These documents shall be listed in the order that they appear in the procedure.

Applicable Documents: List all relevant procedures and forms.

Definitions: All words, acronyms, or phases used in a manner not widely understood.

Overview: Comments of a general nature, detailing the approach, objectives or general guidelines.

General Requirements: List requirements that apply to all affected personnel. An example would be:

There will only be one change allowed for each engineering change package.

Procedure: A detailed step-by-step statement defining specific tasks to be performed and identifying who has the responsibility for performing them.

3.9.0   EXAMPLE ENGINEERING PROCEDURE

See Appendix C for another example of an engineering procedure.

⊕	**Engineering Procedure**		
Title: Part Numbering System	Eng. Proc. No.:		EP-02-05
Release Date: 04/15/00	Revision: A	Page: 1 of 6	

1.0   Purpose

The purpose of this procedure is to provide instructions, and to assign responsibilities for assigning new part numbers to items that are inventoried (parts, subassemblies or assemblies). It will also provide instructions for preparing and submitting the Part Number request form.

2.0   Applicable Documents

Engineering Procedure:
    EP-2-2 Document Number Assignment Logbook

Engineering Form:
    E023 Part Number Request

3.0   Definitions

Buy Part/Assembly: A part or assembly which is made uniquely for the company by an outside vendor to the company's unique design specifications, that are supplied to the vendor in the form of engineering drawing(s) and/or engineering specification(s) accompanied by a bill of material (if it is an assembly).

Prepared By:	
Engineer:	Date:

Approved By:	
Engineering Manager:	Date:
Manufacturing Manager:	Date:
Quality Manager:	Date:

Example–Engineering Procedure Page 1

Engineering Procedure		
Title: Part Numbering System	Eng. Proc. No.: EP-02-05	
Release Date: 04/15/00	Revision: A	Page: 2 of 6

3.0   Definitions (Cont'd)

Standard Part: An off-the-shelf item which is available commercially and is not made to any specifications other than those of the vendor.

Part Number: A number which uniquely identifies a piece part, subassembly or assembly. For make-or-buy parts, the part number consists of the base number plus the dash number.

Base Number: A five-digit number included in the part number of a make or buy part which identifies it as a unique design part. The number is assigned by the company to a new part, its engineering drawing and bill of material. It includes any zeros added to the left side of the number which are needed to make it a five-digit number.

Dash Number: The last two digits of a make or buy part number. The dash number changes whenever a non-interchangeable change is made to the part. The dash number shall appear in all documentation. When a new part number is assigned, the dash number shall always begin with –00.

Revision Level: The revision level is a one or two alpha character which represent changes to an engineering drawing. The revision level is not part of the part number. It changes each time the engineering drawing changes (whether the actual part changes or not). The main purpose of the revision level is to account for changes to the engineering drawing. Another purpose of the revision level, particularly at the assembly level, is to account for non-interchangeable changes to a part within a bill of material. Whenever a non-interchangeable change is made to a piece part, its assembly part number must either change or receive a revision level change.

4.0   General Requirements

4.1   All part numbers for company parts used in production contain five digits. For those production part numbers that contain less than five digits, zeros must be added to the left of the part number to make it a five-digit number. This helps align the numbers for easy sorting.

⊕	**Engineering Procedure**	
Title: Part Numbering System		Eng. Proc. No.: EP-02-05
Release Date: 04/15/00	Revision: A	Page: 3 of 6

4.2   Because any zeros added to the beginning of a company part number becomes an integral part of that part number, they shall appear on all company documentation representing that part number.

5.0   Part Number Procedure

This procedure is divided into the following parts:

- Part Number Procedure

- How to Prepare the Part Number Request form

5.1   *Part Numbering System.* The following is an example of a typical company make-or-buy part and revision level as it appears in the title block of an engineering drawing.

Base Number	Dash Number	Revision Level
00221	-06	B

Actual part number and revision: 00221-06B

5.1.1   *Base Number.* The base number for a typical make-or-buy part is always 5 digits long. A base number for a part will always remain the same, regardless of the type of change. However, if the dash number changes more than 99 times, a new base number must be assigned.

The base number system outlined above applies to all three product phases: Development, Preproduction and Production

5.1.2   *Dash Number.* The dash number always consists of two numeric digits. Upon initial assignment of a new part number, the dash number is always –00.

Example–Engineering Procedure Page 3

(🌐) **Engineering Procedure**		
Title: Part Numbering System	Eng. Proc. No.: EP-02-05	
Release Date: 04/15/00	Revision: A	Page: 4 of 6

5.0   Part Number Procedure (Cont'd)

5.1.3   *Non-Interchangeable Change.* Whenever a non-interchange-able change is made to a part, the dash number changes by one digit. This can happen up to 99 times, as outlined in Paragraph 5.1.1.

Example: -01, -02, -03…-99

5.2   *Interchangeable Change.* When an interchangeable change is made to the part, the dash number remains unchanged. This means that there can be two different looking parts that can be stored in the same bin, because they are interchangeable.

5.3   *Document Revision Level.* The document revision level consists of a one or two digit alphabetic representation. The document revision level is the only thing that must change every time a change is made to an engineering drawing.

The engineering drawing revision level appears in the engineering drawing revision block, in the top right hand corner, and at the bottom of the drawing next to the part number. It also appears in the Document Master List database for each part number.

5.4   *Revision Levels During Product Phases.* Engineering drawing revision levels as they apply to different product phases are detailed below.

5.4.1   *Development Phase.* During the Development phase, document revision levels shall be designated P1 (for development release), P2, P3, etc.

5.4.2   *Preproduction Phase.* When the engineering drawing is released into the Preproduction phase, the revision level shall become A (for Preproduction release), And subsequent changes shall be designated B, C, etc. The Development revision level history shall be erased from the drawing when it is released to Preproduction.

5.4.3   *Production Phase.* When the engineering drawing is released into the Production phase, the revision level shall continue on as D, E, etc. with the next change after release. The revision level shall not change at the time of production release.

Example–Engineering Procedure Page 4

( )	**Engineering Procedure**	
Title: Part Numbering System	Eng. Proc. No.: EP-02-05	
Release Date: 04/15/00	Revision: A	Page: 5 of 6

The following procedure describes who is responsible and what they are supposed to do for each processing step.

5.5   Engineer

Steps

1. Prepare the Part Number Request form E023. Generate a new part number using the above part numbering system in Paragraph 5.0.

2. Forward the completed Part Number request form to the Engineering Manager for review and approval.

5.6   Engineering Manager

3. Verify that the Part Number Request form is complete and correct. If approved, sign and date, then return the form to the Engineer for further processing.

5.7   Engineer

4. Forward the approved Part Number Request form to engineering document control for the assignment of part numbers.

5.8   Engineering Document Control

5. Enter the next available number in the logbook and on the Part Number Request form.

6. Forward a copy of the Part Number Request form to the Engineer.

7. Run copies, stamp, distribute, and then file the original in the New Part Number Request file by part number.

Example–Engineering Procedure Page 5

(🜨)	**Engineering Procedure**	
Title: Part Numbering System	Eng. Proc. No.: EP-02-05	
Release Date: 04/15/00	Revision: A	Page: 6 of 6

6.0 Overview

The procedure for processing the Part Number Request form shall be followed by each individual responsible for entering information on the form. Each circled number below corresponds to the circled number on the following Part Number Request form E023.

6.1 Engineer

❶ Enter your name and date that the Part Number Request form was completed.

6.2 Engineering Manager

❷ Enter your name and date upon approval of the information entered on the form.

6.3 Engineering Document Control

❸ Assign the next available part number from the Document Assignment Logbook, then enter the number in the Part Number Request form E023.

6.4 Engineer

❹ Enter the part description.

❺ Enter the manufacturer's part number, if this is a purchased part.

Example–Engineering Procedure Page 6

(🌐)	**Part Number Request**		
Engineer: ❶			Date:
Engineering Manager: ❷			Date:
Part Number ❸	Description ❹		Manufacturer's Part Number (If Applicable) ❺

Form E023 (Procedure EP-2-5) 8/2/01

Example–Part Number Request

3.10.0 FORMS

When engineering procedures require forms, they should follow a similar format for ease of use. The forms should be nice to look at and should follow a systematic flow of thought. The form has to be an integral part of the procedure. Forms cannot stand alone and there has to be a written procedure to explain how to use them. A form could be considered a template with fill-in-the-blank entry boxes or a checklist of answers to circle.

According to the American Heritage Dictionary, a *form* is *1) A document with blanks for the insertion of details or information.*

A form that is to be used with an engineering procedure is a *document.* The form instructs the end user about details required in preparing test report, for an example. Once the form is filled out, however, it becomes a *record*, a written statement of facts pertaining to a specific event, a test that took place at a specific date, time, and location.

Forms consist of documented evidence that an operation complies with the system established by the first three tiers of documentation. For example, employees might fill out a checklist to indicate that they carried out a job according to their work instruction documentation.

Forms are used to:

- Request
- Analyze
- Change
- Notify
- Record
- Report
- Input into computer
- Log

Following are some examples of engineering operations where forms could be used to capture information.

- Product Introduction
- Engineering Change Proposal
- Cost Analysis Report
- Specification Change Notice
- Bill of Material Input

3.11.0 FORM FORMAT

The form is used to capture useful information and approval signatures. This section shows how to develop a complete form. The form is used to convey the associated engineering change number, what models were affected, and who prepared and approved it. With a form, you will be answering the following questions:

- Who prepared the form?

- What is the associated engineering change number?

- What models were affected?

- What is the priority of the change?

- What is the part number and description?

- What is the vendor disposition?

- What is the reason for the change?

- What is the justification/impact?

- Who reviewed and approved the form?

3.12.0 EXAMPLE FORM

Here is an example form of an Approved Vendor List.

🌐	**Approved Vendor List**

Engineer: ❶	Date:	Associated EC No.: ❷

Model(s) Affected: ❸	Priority: ❹ __ Routine __ Urgent

Part Number: ❺	Description:

Vendor Disposition: ❻

_ Add   Vendor Name _____ Vendor Part No. _____

_ Change   Vendor Name _____ Vendor Part No. _____

_ Delete   Vendor Name _____ Vendor Part No. _____

Disposition (for delete only) ❼

Use-as-is _____   Rework _____   Scrap _____

Reason for Change:
  ❽

Justification/Impact:
  ❾

Authorizations:
  ❿

Engineering:	Date:
Manufacturing:	Date:
Quality:	Date:
Purchasing:	Date:

Form E035 (Procedure EP-5-4)

Example–Approved Vendor List

Approved Vendor List Form Preparation. Each of the following circled numbers corresponds to the circled numbers on the example approved vendor list form.

Engineer

❶ Originator: Enter your name and date when the form was completed.

❷ Associated EC No.: Enter the associated Engineering Change number, if applicable.

❸ Models Affected: Enter the model(s) affected by the change.

❹ Priority: Check the priority of the change. ("Urgent" is for line down situations only.)

❺ Part Number: Enter the part number and description for which the change is being requested.

❻ Vendor Disposition: Check the action that is being requested for the reference part number. (All three boxes may be checked.)

❼ Disposition: Only used if the "Deleted" box is checked. Enter the disposition of parts being deleted.

❽ Reason for Change: Enter the reason for the change request, e.g., additional source, added per an engineering change.

❾ Justification/Impact: Enter the justification for why the change should be made, including the impact on production or products.

Managers

❿ Authorizations: Enter your name and date, if approved.

4

Engineering Procedure
Manuals

Manuals are a tried and true method of having procedural information available for end users. This section is for a manual paper-based documentation system; therefore, hard copies of procedures will be three-hole punched and placed in binders and distributed throughout the company. When engineering procedures are released for use, they will be distributed to all engineering procedure manuals. As changes are made, the down-level procedures are obsoleted and retained for history.

The engineering department staff is responsible for approving, publishing, and distributing the engineering procedure manuals and any subsequent revisions. The manual will be reviewed and updated annually or as required to maintain an effective engineering program. The manual is maintained on a controlled copy basis, with manual holders receiving copies of new or revised pages as they are issued and notices to remove deleted pages. Generally, only manuals that have identification numbers will be controlled and maintained.

There will need to be periodic, random audits performed to determine if manual holders are properly maintaining their individual copies. All changes to the contents of the manual are made through the engineering department.

4.1.0   MANUAL COMPONENTS

Engineering procedure manuals will follow the customary format of a cover page, spine, 3-ring binder, dividers, introduction and
maintenance page, and a table of contents with revision history page as
shown in Fig. 4.1.

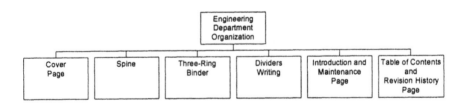

Figure 4.1. Engineering procedure manual components.

4.1.1   Cover Page

The manual will have a cover (Fig. 4.2) that states that it is an engineering procedure manual and there will be an unique identification number, the name of the person to whom the manual is assigned, and the date the manual was assigned.

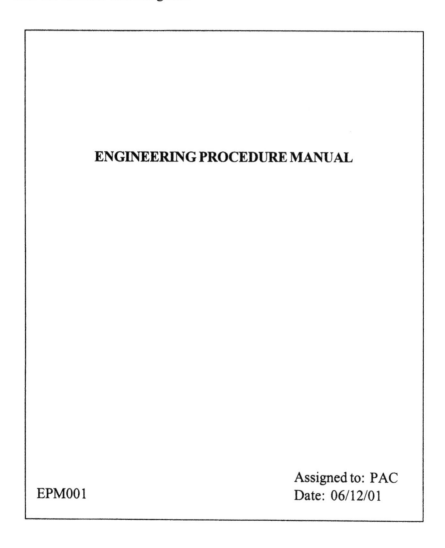

ENGINEERING PROCEDURE MANUAL

Assigned to: PAC

EPM001   Date: 06/12/01

Figure 4.2. Engineering procedure manual cover page.

4.1.2 Spine

There are several ways that you can print the information on the spine of a manual as shown in Fig. 4.3.

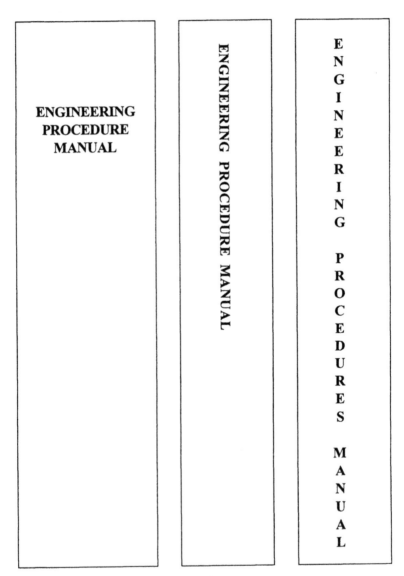

Figure 4.3. Manual spine information format.

4.1.3   Binders

It is a good idea to start with a binder that is about half full. This will allow ample room for expansion and will not look as empty in the beginning. Generally, a 2" binder is the maximum desirable size. The 2" dimension is for inside diameter of the rings, not the size of the binder's spine. Of course the number of engineering procedures that you plan on producing will determine the size of the manual.

For most manuals, a 3-ring binder is preferable. If you have a lot of material to include in your manual, you should consider using a D-ring binder. These binders will hold more paper then a round-ring binder. Since the D-rings are mounted on the back cover of the binder, the binder lies flat and the paper does not move when the binder is opened.

The front, spine and back cover should have plastic jackets where paper or card stock can be inserted with the manual's information printed on them.

4.1.4   Dividers

The divider should be made out of material that will last the lifetime of the manual. The tabs on the dividers should be printed on the front and back with the word "Section" and the section number only. Figure 4.4 illustrates the suggested sections of an engineering procedure manual.

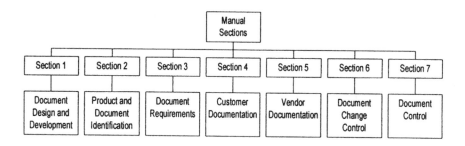

Figure 4.4. Engineering procedure manual sections.

4.2.0   ENGINEERING PROCEDURE MANUAL PAGE

There needs to be several manual pages, such as introduction and table of contents. These pages are used for recording the manual's number, release date and the page numbers. With this list you will be answering the following questions:

What is the manual's number?

When was the manual released?

What is the page number?

Engineering Procedure Manual Page - Preparation. Each of the following circled numbers corresponds to the circled numbers on the example engineering procedure manual page.

❶ Manual Number: Enter the engineering procedure manual number. (See Ch. 6 under Engineering Procedure Manual Numbers.)

Example: EPM001

❷ Release Date: Enter the date that the manual was released for use.

❸ Page: Enter the page number.

Example: Page 1 of 3. This method is best for change control purposes.

⊕	Engineering Procedures Manual	
Manual Number: ❶	Release Date: ❷	Page: ❸

Example–Engineering Procedures Manual

4.2.1 Manual Introduction and Maintenance Page

(🌐)	Engineering Procedures Manual	
Manual Number: EPM001		Release Date: 04/10/03
		Page: 1 of 5

Introduction

This manual contains procedures that are to be followed by all engineering personnel. It will serve as a permanent reference and working guide for engineering personnel in the day-to-day administration of their work duties.

These written procedures should increase understanding and help assure uniformity throughout the department. It is the responsibility of each member of management to administer these procedures in a consistent and impartial manner. For a more complete understanding, employees shall read all of the procedures in detail.

This manual is organized into sections, each dealing with a particular topic. The sections are outlined in the table of contents. Within each section of this manual, cross references are indicated. Each section contains information specific to the engineering department. As mentioned, this manual gives steps to be followed as well as samples and/or forms related to specific tasks within engineering.

Manual Maintenance

Recommendations for additions to, deletions from, and changes to this manual should be addressed to the engineering department.

Example–Engineering Procedures Manual Page 1

🌐 **Engineering Procedures Manual**	
Manual Number: EPM001	Release Date: 04/10/03
	Page: 2 of 5

Manual Maintenance (Cont'd)

Changes to this manual will be made by page revision of each procedure within the manual. New page changes will be inserted into the manual by the end user. Procedures will become effective upon approval by the manager of engineering.

All holders of this manual who are on automatic distribution will receive all changes. All manual holders are responsible for maintaining their assigned copies. Changes or additions should be immediately inserted into the manual.

If the manual is checked out in your name, you are responsible for it. It is imperative that under no circumstances is the manual to be given to anyone else. If you no longer need the manual, return it to engineering document control so you may be removed from the distribution list.

Example–Engineering Procedures Manual Page 2

4.2.2   Table of Contents with Revision History Page

🜨	**Engineering Procedures Manual**	
Manual Number: EPM001		Release Date: 04/10/03
		Page: 3 of 5

Table of Contents

Section 1: Product Design and Development

		Revision
EP-01-01	Product Planning	A
EP-01-02	Product Introduction	C
EP-01-03	Make-or-Buy Committee	etc.
EP-01-04	Product Safety	
EP-01-05	Engineering Review Board	
EP-01-06	Product Phases	
EP-01-07	Design Reviews	
EP-01-08	Baseline Management	
EP-01-09	Product Structuring	
EP-01-10	Compatibility Books	
EP-01-11	Drawing Trees	

Section 2: Product and Documentation Identification

EP-02-01	Document Numbering System
EP-02-02	Document Number Assignment Logbook
EP-02-03	Block Number Assignment
EP-02-04	Model Number Assignment
EP-02-05	Part Numbering System
EP-02-06	Part Marking System
EP-02-07	Serial Numbering System
EP-02-08	Item Master Input
EP-02-09	Bill of Material
EP-02-10	Where-Used Input

Example–Engineering Procedures Manual Page 3

⊕	**Engineering Procedures Manual**
Manual Number: EPM001	Release Date: 04/10/03
	Page: 4 of 5

Table of Contents (Cont'd.)

Section 3: Documentation Requirements

EP-03-01   Engineering Documentation Types

EP-03-02   Document Flow Chart

EP-03-03   Documentation Order of Precedence

EP-03-04   Engineering Lab Notebook

EP-03-05   Product Specification

EP-03-06   Configuration Control Document

EP-03-07   Drawing Format Requirements

EP-03-08   Printed Circuit Board Documentation

EP-03-09   Component Part Request

EP-03-10   Deliverable Documentation Review

Section 4: Customer Documentation

EP-04-01   Customer Documentation Control

EP-04-02   Engineering Change Proposal

EP-04-03   Deviation/Waiver

Section 5: Vendor Documentation

EP-05-01   Vendor Documentation Control

EP-05-02   Single Source Authorization

EP-05-03   Limited Buy Authorization

EP-05-04   Approved Vendor List

Example–Engineering Procedures Manual Page 4

(🌐)	Engineering Procedures Manual	
Manual Number: EPM001		Release Date: 04/10/03
		Page: 5 of 5

Table of Contents (Cont'd.)

Section 6: Document Change Control

EP-06-01   Change Control System

EP-06-02   Engineering Change Routing

EP-06-03   Engineering Change Request

EP-06-04   Engineering Change Procedure

EP-06-05   Document Mark-Up for Change Package

EP-06-06   Engineering Change Costing

EP-06-07   Prototype Report

EP-06-08   Change Control Board

EP-06-09   Engineering Change Implementation

EP-06-10   Suffix Engineering Change

EP-06-11   Bill of Material Change

EP-06-12   Document Change Notice

EP-06-13   Specification Change Notice

EP-06-14   Field Instructions

EP-06-15   Equivalent Item Authorization

Section 7: Document Control

EP-07-01   Document Control System

EP-07-02   Document Release

EP-07-03   Document Distribution

EP-07-04   Engineering Change/Part Number Input

EP-03-05   Security Marking of Documents

EP-07-06   Copy Request

EP-07-07   Document Original Withdrawal

EP-07-08   Literature Search/Acquisition

EP-07-09   Forms Control

Example–Engineering Procedures Manual Page 5

4.3.0 ENGINEERING PROCEDURE MANUAL DISTRIBUTION

After initial release of the Engineering Procedure Manual, Engineering Document Control will be responsible for issuing copies to all individuals on the distribution list. The Manual Distribution List is maintained by Engineering Document Control. All department heads and all other personnel directly responsible for the activities described in the manual will be listed on the Manual Distribution List. Copies of the manual may be distributed for information only to individuals not on the distribution list upon request. All such copies will be clearly marked on the cover page (Fig. 4.5), "NOT A CONTROLLED COPY."

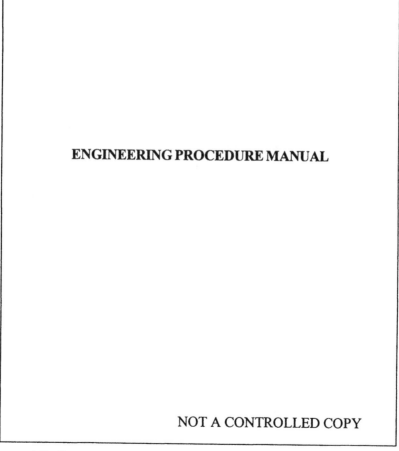

Figure 4.5. Cover page.

A unique identifying number will be assigned to each controlled copy of the manual. The identifying number shall appear on the title page of all controlled copies. The same control number will appear adjacent to the name of each manual recipient on the distribution list.

4.3.1   Distribution of Manuals to Individuals

Here is an example of a memo (Fig. 4.6) that can be used to distribute manuals to individuals.

4.3.2   Distribution of Manuals to Departments

Here is an example of a memo (Fig. 4.7) that can be used to distribute manuals to departments.

4.3.3   Document Agreement Form

The manual has been developed for the purpose of and use by company employees. Due to the confidentially of the contents of this manual, the following requirement has been established:

- Each person receiving a copy of the manual is required to sign a Controlled Document Agreement Form.

A Controlled Document Agreement Form (Fig. 4.8) has been established to control manual copies. By signing the form, an employee agrees not to copy the confidential information contained in the manual or remove it from the company.

MEMO

To:

From:

Subject: Manual Distribution

This engineering procedure manual has been issued to you. It has been assigned a unique number for accountability and change control purposes. If you no longer need the manual, return it to engineering document control so your name can be removed from the distribution list.

The manual is distributed and maintained on a controlled-copy bases. You will be receiving copies of new or revised pages as they are issued, as well as notices to remove deleted procedures. There will be periodic audits to ensure that the manual is complete and up-to-date.

If anyone within the company should need a copy of the manual, please contact engineering document control.

Distribution:

Manual Number:_____

Issued to: _____

Date: _____

Figure 4.6. Manual distribution to individuals memo.

MEMO

To:

From:

Subject: Manual Distribution

This engineering procedure manual has been issued to your department. As manager, you are responsible for the manual and all updates will be sent to you. It has been assigned a unique number for accountability and change control purposes. If you no longer need the manual, return it to engineering document control so your name can be removed from the distribution list.

The manual is distributed and maintained on a controlled-copy bases. You will be receiving copies of new or revised pages as they are issued, as well as notices to remove deleted procedures. There will be periodic audits to ensure that the manual is complete and up-to-date.

If anyone within the company should need a copy of the manual, please contact engineering document control.

Distribution:

Manual Number: _____

Issued to: _____

Date: _____

Figure 4.7. Manual distribution to departments memo.

CONTROLLED DOCUMENT AGREEMENT

The contents of the manual has been specifically organized to assist those individuals assigned and employed within the company to perform their functions in an acceptable manner and in concert with other functions. The manual has been developed strictly and solely for the purpose of and use by employees of the company, and any disclosure, plagiarizing, or other use made of the structuring or contents is prohibited.

I, the undersigned, to whom this manual is assigned, fully understand and agree that I will use it for its intended purpose, will maintain it in a current and reasonable condition, and, in the event of my termination, will be held accountable for it and stand ready to forfeit it when asked to do so.

Manual Number:_____

Signed:_____

Date: _____

Form E100 (Procedure EP-7-3) 7/10/00

Figure 4.8. Controlled Document Agreement.

4.3.4   Engineering Procedure Manual Distribution List

A user manual distribution list will need to be developed. The list documents the number of manuals copied and distributed to company employees. By checking the list you can identify which employee has a particular manual in their possession, the date they received the copy, and whether or not that copy has been returned. With this list you will be answering the following questions:

What is the manual's number?

When was the manual released?

What is the page number?

Engineering Procedure Manual Page - Preparation. Each of the following circled numbers corresponds to the circled numbers on the example engineering procedure manual page.

❶ Manual Number: Enter the engineering procedure manual number. (See Ch. 6 under Engineering Procedure Manual Numbers.)
Example: EPM001

❷ Issued to: Enter the name of the person that the manual was issued to.

❸ Date Issued: Enter the date that the manual was issued.

❹ Date Returned: Enter the date that the manual was returned.

Engineering Procedure Manual Distribution List			
Manual Number	Issued to:	Date Issued	Date Returned
❶	❷	❸	❹

Example–Manual Distribution List

4.3.5   Example—Engineering Procedure Manual Distribution List

🌐 Engineering Procedure Manual Distribution List			
Manual Number	Issued to:	Date Issued	Date Returned
EPM001			
EPM002			
EPM003			
EPM004			
EPM005			
EPM006			
EPM007			
EPM008			
EPM009			
EPM010			
EPM011			
EPM012			
EPM013			
EPM014			
EPM015			
EPM016			
EPM017			
EPM018			

Example–Manual Distribution List (with sample entries)

4.4.0   MANUAL REVISIONS

The engineering manager must approve all revisions to the manual. The latest revision level (letter) and the date of the revision shall be shown in the title block on each procedure in the manual and in the table of contents. Upon approval, engineering document control will distribute copies of revisions to all individuals on the distribution list. Revisions will be distributed by hand carrying. Revised copies will be collected and destroyed.

Here is an example of a memo (Fig. 4.9) that can be used to distribute changed pages to manual holders.

MEMO

To:

From:

Subject: Changed Pages Distribution

The attached procedure and table of contents has been released and is now in effect. Insert it into your engineering procedure manual in the proper place. Inform your employees who are affected by the procedure of its release. As always, the attached table of contents replaces the existing one. If you have any questions regarding this or any other engineering procedure, contact engineering document control.

EP-06-03   Engineering Change Request   Revision C

Revised Table of Contents

Distribution:

Manual Number:_____

Issued to: _____

Date: _____

Figure 4.9. Changed pages distribution.

5

Change Control

Where does it say that you have to change, review, approve, and follow engineering procedures? Well, if you are or want to be in compliance with ISO 9000 regulations then the following will apply to you.

ISO 9000 Sec. 4.5, Document Control; Subsection 4.5.2, Document Changes/Modifications:

Changes to documents shall be reviewed and approved by the same organizations that performed the original review and approval unless specified otherwise. The designated organizations shall have access to pertinent background information upon which to base their review and approval.

Where practicable, the nature of the change shall be identified in the document or appropriate attachments.

A master list or equivalent document control procedure shall be established to identify the current revision of documents in order to preclude the use of non-applicable documents.

Documents shall be re-issued after a practical number of changes have been made.

This chapter covers the method used for document review and approval, how to request changes to released documentation, how the document review board operates, and how to record the revision history of changes to all of the documentation.

5.1.0 DOCUMENT REVIEW AND APPROVAL

Now that documentation has been written, it is time for review and approval by qualified individuals. Your document must be reviewed by the engineering department and then by outside departments, if required. Review and approval will be required for each of the following engineering documents: policies, departmental instructions and engineering procedures and all changes to these documents. All of these documents will need to go through the Document Review Board for review and approval. The Document Review Board is explained later on in this chapter. Figure 5.1 illustrates the flow of all documents for review and approval.

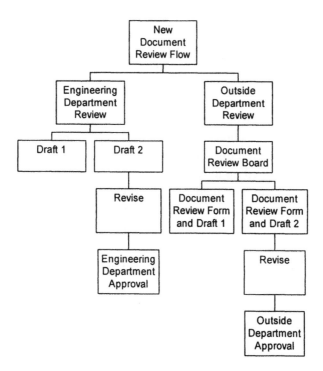

Figure 5.1. New document review flow.

5.1.1   Engineering Department Review

After you have finished writing your documents, you will need to distribute them for review and approval by the engineering department. You do not need the Document Review form for this step. Your documents will go through several iterations (as shown above) within the engineering department. Stamp the first review copy '"Draft 1;" this way the reviewers will know from the front of the document what part of the review cycle the document is in. When your documents come in from review, you can tell at a glance the state of the document. When the document returns with the reviewers comments marked up in red, incorporate the changes and distribute "Draft 2," and so forth until there are no more changes required. This allows you to have all of your ducks in a row before distributing your work to outside departments. Do not sign the document at this time because this is just a draft copy, not the final.

5.1.2   Outside Department Review and Approval

Next, you will prepare a document review form (see below) and attach a copy of the document to be reviewed (Fig. 5.1), and then distribute the document review form and document simultaneously to the Document Review Board members for review or approval. Give the reviewers a week or two to review the document, depending on the magnitude of the document. Inform the reviewers that they can return their comments to you or bring them to the next Document Review Board meeting. You will go through the same iterations with the outside departments as you did with your own. Call the first distribution copy "Draft 1." When you get the document back marked up in red, incorporate the changes and distribute "Draft 2" and so forth until there are no more changes required.

After you receive the marked-up copies from each reviewer the Document Review Form will be marked with either:

> *"Approved."* The document is ready for use.

> *"Conditionally Approved."* You will need to show
> that the conditions have been met.

"Not approved." You will need to call a Document Review Board meeting to resolve any issues. Also, when you get different department inputs that conflict with each other, you will need to call a Document Review Board meeting to resolve those issues.

After all concerns have been addressed and changes incorporated, have your own department verify that you have incorporated all of the changes correctly. Next, sign and date the document, then have your manager sign and date the document and bring it to the next Document Review Board meeting. There, all board members will sign the document. After approval the document will be released into engineering document control where it will be readied for distribution to manuals.

5.1.3   Document Review Form

With this form you will be answering the following questions:

Who was the originator of the form and when did they check out the number?

What is the document review form number?

What is the document number and title?

Which type of document is being reviewed?

What was the reviewer's response?

Who was the reviewer?

Who were copies distributed to?

	Document Review Form		
Originator: ❶	Date:		DRF No.: ❷
Document Number: ❸	Title:		

Documentation Types: ❹

_____ Policy   _____ Departmental Instruction
          _____ Engineering Procedure

Reviewer Response: ❺

>   Mark your comments directly on the document.
> After your review is complete, either hold on to
> your mark-ups for the next Document Review
> Board meeting or return them to Engineering.
> Engineering will bring all marked-up copies to
> the next meeting for discussion and agreement.

     Your comments are due by: _____. If more time is
needed, contact engineering for an extension.

_____ Approved

_____ Conditionally Approved (See attached comments.)

_____ Not Approved, explain.

Reviewer Signature: ❻   Date:

Distribution: ❼

Engineering:

Manufacturing:

Quality:

Form E001 (Procedure EP-1-5) 7/12/01

Example-Document Review Form.

Document Review Form Preparation. Each circled number below corresponds to the circled number in the example document review form.

❶   Originator: Enter your name and date when the form is initiated.

❷   DRF No.: Enter the document approval form number. (See Ch. 6 under Document Review Form Numbers.)

❸   Document Number: Enter the document number and title.

❹   Documentation Types: Mark the appropriate box for the document that is being reviewed and approved.

❺   Reviewer Response: Mark the appropriate response after your review.

❻   Reviewer Signature: Sign and date when your review is complete.

❼   Distribution: Enter distribution names. This is a standard distribution for all documents.

5.1.4 Example Document Review Form

⊕ **Document Review Form**	

Originator:	Date: 6/1/01	DRF No.: 015
Document Number: DI-01	Title: Where-Used Input	

Documentation Types:

_____Policy   ●_____Departmental Instruction
     _____ Engineering Procedure

Reviewer Response:

      Mark your comments directly on the document.
After your review is complete, either hold on to
your mark-ups for the next Document Review
Board meeting or return them to Engineering.
Engineering will bring all marked-up copies to
the next meeting for discussion and agreement.

      Your comments are due by:__06/12/01_. If more time is
needed, contact engineering for an extension.

_____Approved

●_____Conditionally Approved (See attached comments.)

_____ Not Approved, explain.

Reviewer Signature:	Date:

Distribution:

Engineering:

Manufacturing:

Quality:

5.2.0 CHANGE CONTROL METHODS

The change control process starts with a request to make changes to released documents. (See Fig. 5.2.) Anyone can request a change by preparing a Document Change Request form and attaching a marked up copy of the document that needs changing. The Document Change Request form will be distributed to the Document Review Board members for review. At the next Document Review Board meeting the decision will be made to reject or accept the change request.

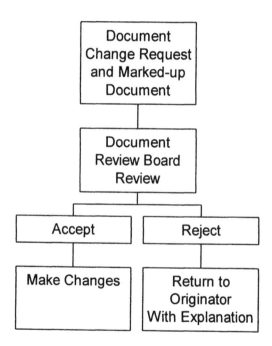

Figure 5.2. Document change request method.

5.2.1   Document Change Request Form

This section will show you how to complete a document change request form. The method of preparation is the same for all departments. Document change requests will be reviewed and approved by the Document Review Board. With the document change request form you will be answering the following questions:

- Who was the originator of the form and when was it originated?

- What is the document change request form number?

- What type of document is being changed?

- What is the document's disposition?

- What is the priority of the change request?

- What are the document's number, revision, and title?

- What is the reason for the change?

- What is the justification for the change?

- Who reviewed and approved the change request?

☻	**Document Change Request Form**

Originator: ❶	Date:	DCRF No.: ❷

Document Type: ❸	Disposition: ❹
_____ Policy	____Change
_____ Departmental Instruction	____Obsolete
_____ Engineering Procedure	Priority: ❺
_____ Form	____Routine ____Urgent

Document Affected:

Document Number	Revision	Title
❻		

Reason for change:

❼

Justification:

❽

❾ Approved By:

Engineering Manager:	Date:
Manufacturing Manager:	Date:
Quality Manager:	Date:

Form E002 (Procedure EP-6-3) 8/19/01

Example–Document Change Request Form

Document Change Request Form Preparation. Each circled number below corresponds to the circled number in the example document change request form.

❶ Originator: Enter your name and date when the form is initiated.

❷ DCRF No.: Enter the document change request number. (See Ch. 6 under Document Change Request Form Numbers.)

❸ Document Type: Mark the appropriate box for the document that is affected.

❹ Disposition: Mark the appropriate box depending on the document's disposition.

❺ Priority: Mark the appropriate box depending on the priority of the change.

❻ Document Number, Revision, and Title: Enter the document's number, revision, and title.

❼ Reason for Change: Enter the reason and description of the change or attach marked up documents.

❽ Justification: Enter the justification for the change.

❾ Approved By: Enter your name and date when the change request form is complete and ready for processing.

5.2.2   Example—Document Change Request Form

(🌐)	**Document Change Request Form**

Originator:	Date:	DCRF No.: 003

Document Type:	Disposition:
_____ Policy	_●_Change
_____ Departmental Instruction	___Obsolete
_●_ Engineering Procedure	Priority:
_____ Form	_●_Routine___Urgent

Document Affected:

Document Number	Revision	Title
Ep-02-05	B	Part Numbering System

Reason for change:

The form number E044 on page 6 should be E023.

Justification:

There would be lost time and money after finding out that you have the wrong form.

Approved By:

Engineering Manager:	Date:

Manufacturing Manager:	Date:

Quality Manager:	Date:

Form E002 (Procedure EP-6-3) 8/19/01

Example–Document Change Request Form (with sample entries)

5.3.0   DOCUMENT REVIEW BOARD

Establish a review board with representation from key personnel (Fig. 5.3). A good change control system is necessary to make sure that previous documentation efforts have not been lost. Change control starts when policies, departmental instructions, and engineering procedures have been approved and released into engineering document control. From this time on every change to any of the documents must go through the Document Review Board for review and approval.

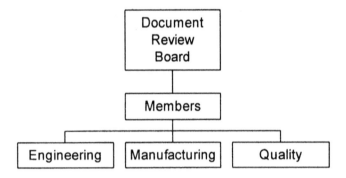

Figure 5.3 Document review board.

The Document Review Board is responsible for the review and approval of all policies, departmental instructions, engineering procedures, and forms. The Document Review Board will receive documents from engineering procedure writers, and are required to be accurate and efficient in their reviews. A specific length of time and completion date will be assigned to each document.

The Document Review Board:

> ➤ Convenes once a week or sooner as required.

> ➤ Reviews and either accepts or rejects each proposed new, revised, or obsoleted policy, departmental instruction, engineering procedure, or form.

> ➤ Determines if a document is a policy, departmental instruction, or engineering procedure.

> ➤ Reviews documents line-for-line and reaches an agreement on which changes need to be incorporated.

> ➤ Establishes an effectivity date for when the changes will be incorporated into the system.

After the Document Review Board has been operating, you can create a memo (Fig. 5.4) that notifies the members about which document is going to be reviewed and approved at the next Document Review Board meeting.

5.4.0 REVISION HISTORY SHEET

There will need to be a revision history sheet created for each policy, departmental instruction, and engineering procedure. The revision history sheet starts when a document is released for use. This method keeps a running history of the changes to each document. There will be lots of times where you will be referring back to when a certain change took place. The initial release will be 'A' and the second 'B' and so-forth. Some companies start with the revision system as 'No Change' or the revision box is left blank. All of these will work but you have to agree on one method or later on you will be sorry. With this sheet you will be answering the following questions:

- What is the document's number?
- Where is the revision letter?
- What is the document's title?
- Who filled out the sheet?
- What was the date that the sheet was filled out?

Up-Coming Document Review Board Meeting Memo

MEMO

To:

From:

Subject: Document Review Board

> The next Document Review Board meeting will be 3/10/01. If you are unable to attend send someone in your place who can speak and sign for you.

> The agenda for the meeting will be as follows. Bring your marked up documents to the meeting for discussion.

- Documents to be discussed and approved this meeting:

    EP-02-02 Document Numbering System

    EP-03-04 Engineering Logbooks

- Distribution of first drafts for your review.

    EP-02-03 Bill of Material

Distribution:

Figure 5.4. Document review board meeting memo.

🌐		Revision History Sheet		
Document Number	Rev.	Title	Initials	Date
❶	❷	❸	❹	❺

Example–Revision History Sheet

Revision History Sheet - Preparation

Each of the following circled numbers corresponds to the circled numbers on the example revision history sheet.

❶ Document Number: Enter the document number.

❷ Rev.: Enter the revision letter of the document.

❸ Title: Enter the document's title.

❹ Initials: Enter your initials after logging in the information.

❺ Date: Enter the date after the information is loaded onto the sheet.

5.5.0   REVISION HISTORY LIST

There will need to be a revision history list to keep a running history of all of the changes to each policy, departmental instruction, and engineering procedure. With this list you will be answering the following questions:

- What is the document's number?
- Where is the revision letter?
- What is the document's title?
- Who filled out the list?
- How many pages are there?
- What is the document change request form number?
- Who logged in the information?
- When was the information entered?

⊕	**Revision History Sheet**			
Document Number	Rev.	Title	Initials	Date
EP-02-05	A	Initial release		
EP-02-05	B	The form number E044 on page 6 was changed to E023		
etc.				

Example–Revision History Sheet (with sample entries)

✪ Revision History List						
Document Number	Rev.	Title	# of Pages	DCRF No.	Initials	Date
❶	❷	❸	❹	❺	❻	❼

Example–Revision History List

Revision History List - Preparation

Each of the following circled numbers corresponds to the circled numbers on the example Revision History List.

❶ Document Number: Enter the document number.

❷ Rev.: Enter the revision letter of the document.

❸ Title: Enter the document's title.

❹ # of Pages: Enter the total number of pages.

❺ DCRF No.: Enter the associated DCRF Number.

❻ Initials: Enter your initials after the information is logged.

❼ Date: Enter the date after the information is logged.

5.5.1   Example—Revision History List

Following is a completed example of a Revision History List.

🜨 Revision History List						
Document Number	Rev.	Title	# of Pages	DCRF No.	Initials	Date
D-01	A	Drawing Number Assignment	6	025		
D-02	A	Where-Used Input Report	5	015		
D-02	B	Where-Used Input Report	5	024		
D-02	C	Where-Used Input Report	5	033		
etc.						
EP-01-01	A	Product Planning	5	010		
EP-01-01	B	Product Planning	5	015		
EP-01-02	A	Product Phases	4	009		
etc.						
P-01	A	Product Documentation	1	027		
P-01	B	Product Documentation	1	040		
P-02	A	Product Phases	2	023		
P-02	B	Product Phases	2	027		
P-02	C	Product Phases	2	030		
etc.						

Example–Revision History List (with sample entries)

6

Engineering Document Control

This chapter covers several subjects including a block number system, document-numbering systems, document titles, logbooks for policies, departmental instructions, engineering procedures, forms, manuals, document review forms, document change request form, and a document master list.

6.1.0 BLOCK NUMBERING SYSTEM

All of the document numbers in this book came from the following blocks of numbers shown in Table 6.1. These blocks of numbers were established prior to assigning numbers to policies, departmental instructions, engineering procedures, forms, manuals, document review forms, and document change request forms. This method controls the number assignment effort so that there will be no duplicate identification numbers on documentation.

Table 6.1. Block Numbering System

Policies
P-01   thru   P-99

This allows for 99 different policies.

Departmental Instructions
D-01   thru   D-99

This allows for 99 different departmental instructions.

Engineering Procedures
EP-01-01   thru   EP-01-99
EP-02-01   thru   EP-02-99
EP-03-01   thru   EP-03-99
EP-04-01   thru   EP-04-99
EP-05-01   thru   EP-05-99
EP-06-01   thru   EP-06-99
EP-07-01   thru   EP-07-99

This allows for 99 different procedures for each section of the manual.

Forms
E001   thru   E999

This allows for 999 different forms.

Manuals
EPM001   thru   EPM999

This allows for 999 different manuals.

Document Review Forms
DRF001   thru   DRF999

This allows for 999 different document review forms.

Document Change Request Forms
DCRF001   thru   DCRF999

This allows for 999 different document change request forms.

Following are the document numbering systems, titles and logbooks for policies, departmental instructions, engineering procedures, forms, manuals, document review forms, and document change request forms.

6.2.0   POLICY NUMBERS

Policies will need unique identification numbers assigned to them for logging and tracking purposes. Identification numbers are used as locators in spreadsheets and in engineering document control. There are two numbering systems defined in this book. This section is for a manual paper-based documentation system and Ch. 7 is for electronic.

Following is an example of an identification number for a policy.

P-01 = Policy

P   =   One letter document designation for a policy.

-01   =   Two digit number. (This allows for 99 policies.)

Figure 6.1 shows an example of what policy numbers look like in a policy manual for the engineering department.

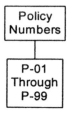

Figure 6.1. Policy numbers.

6.2.1   Example—Policy Numbers and Interpretation

Following are some examples of how to identify a policy by its assigned number. The first letter identifies the type of document, and the next two digits uniquely identify the policy.

Complete Number	Interpretation
P-05	This is the 5th policy written.
P-10	This is the 10th policy written.

6.3.0   POLICY NUMBER ASSIGNMENT

Because the numbering system in this part of the book is for a manual paper-based documentation system, a logbook is required to record each policy number. When assigning new policy numbers, the question is, what is the next available number? A logbook displays the last number that was assigned, and provides a place for a new number to be entered.

6.3.1   Policy Number Assignment Log

Following is an example page from the document number assignment log for policies that is maintained by engineering document control. With this log you will be answering the following questions:

- What is the policy's number?

- What is the title of the policy?

- Who assigned the number?

- When was the number assigned?

![globe logo] Policy Number Assignment Log			
Policy No.	Title	Initials	Date
❶	❷	❸	❹

Example–Policy Number Assignment Log Sheet

Policy Number Assignment Log—Preparation. Each of the following circled numbers corresponds to the circled numbers on the example policy number assignment log.

❶ Policy No.: Enter the policy's number.

❷ Title: Enter the policy's title.

❸ Initials: Enter your initials after logging in the information.

❹ Date: Enter the date after the information is loaded into the log.

6.3.2   Example—Policy Number Assignment Log

Below is an example of a page from a Policy Number Assignment Logbook.

6.4.0   POLICY TITLES

The policy title needs to identify which function is being addressed in such detail that there is no question as to what it is covering.
Policy title examples would be:

Change Control Policy

Customer Documentation Control Policy

Research and Development Policy

Forms Control Policy

⊕	Policy Number Assignment Log		
Policy No.	Title	Initials	Date
P-01	Product Development	PAC	1/2/00
P-02	Product Phases	PAC	2/14/00
P-03	etc.		
P-04			
P-05			
P-06			
P-07			
P-08			
P-09			
P-10			
P-11			
P-12			
P-13			
P-14			
P-15			
P-16			
P-17			
P-18			
P-19			
P-20			
P-21			
P-22			

Example–Policy Number Assignment Log (with sample entries)

6.5.0   DEPARTMENTAL INSTRUCTION NUMBERS

All departmental instructions will need a unique identification numbers assigned to them for tracking purposes. The identification number is used as a locator in spreadsheets and in engineering document control. There are two numbering systems defined in this book. This section is for a manual paper-based documentation system and Ch. 7 is for electronic.

Following is an example of an identification number for a departmental instruction.

D-01 = Departmental Instruction.

D = One letter document designation for a departmental instruction.

-01 = Two digit number. (This allows for 99 departmental instructions.)

Figure 6.2 shows an example of what a departmental instruction's number would look like in a departmental instruction's manual for the engineering department.

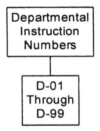

Figure 6.2. Departmental instructions numbers.

6.5.1 Example—Departmental Instruction Numbers and Interpretation

Following are some examples of how to identify departmental instructions by their assigned number. The first letter identifies the type of document, and the next two digits uniquely identify the departmental instructions.

<u>Complete Number</u>   <u>Interpretation</u>

D-04   This is the 4th departmental instruction written.

D-15   This is the 15th departmental instruction written.

6.6.0 DEPARTMENTAL INSTRUCTIONS NUMBER ASSIGNMENT

Because the numbering system in this part of the book is for a manual paper-based documentation system, a logbook is required to record each departmental instruction number. When assigning new departmental instruction numbers, the question is, what is the next available number? A logbook displays the last number that was assigned, and provides a place for a new number to be entered.

6.6.1 Departmental Instruction Number Assignment Log

Following is an example page from the document number assignment log for departmental instructions that is maintained by engineering document control. With this log you will be answering the following questions:

- What is the departmental instruction's number?
- What is the title of the departmental instruction?
- Who assigned the number?
- When was the number assigned?

(🌐) Departmental Instruction Number Assignment Log			
DI No.	Title	Initials	Date
❶	❷	❸	❹

Example–Departmental Instruction Number Assignment Log Sheet

Departmental Instruction Number Assignment Log—Preparation. Each of the following circled numbers corresponds to the circled numbers on the example departmental instruction number assignment log.

❶ DI No.: Enter the departmental instruction's number.

❷ Title: Enter the departmental instruction's title.

❸ Initials: Enter your initials after logging in the information.

❹ Date: Enter the date after the information is loaded into the log.

6.6.2   Example—Departmental Instruction Number Assignment Log

Below is an example of a page from a Departmental Instruction Number Assignment Logbook.

6.7.0   DEPARTMENTAL INSTRUCTION TITLES

The departmental instruction title needs to identify which function is being addressed in such a detail that there is no question as to what it is covering.

Departmental instruction title examples would be:

Master File Update

Drawing Number Assignment

Document Original Control

Where-Used Input Report

Document Release

Document Distribution

🌐 Departmental Instruction Number Assignment Log			
DI No.	Title	Initials	Date
D-01	Drawing Number Assignment	PAC	5/22/00
D-02	Where-Used Input Report	PAC	6/12/00
D-03	etc.		
D-04			
D-05			
D-06			
D-07			
D-08			
D-09			
D-10			
D-11			
D-12			
D-13			
D-14			
D-15			
D-16			
D-17			
D-18			
D-19			
D-20			
D-21			
D-22			

Example–Departmental Instruction Number Assignment Log (with sample entries)

6.8.0   ENGINEERING PROCEDURE NUMBERS

All engineering procedures will need unique identification numbers assigned to them for tracking purposes. Identification numbers are used as locators in spreadsheets and in engineering document control. There are two numbering systems shown in this book. This section is for a manual paper-based documentation system and Ch. 7 is for electronic.

Following is an example of an identification number for an engineering procedure.

EP-01-01 = Engineering Procedure

EP = Two letter document designation for engineering procedure.

-01 = Two digit manual section number. (This allows for 99 manual sections.)

-01 = Three digit procedure number. (This allows for 99 procedures.)

Figure 6.3 shows an example of what the engineering procedure numbers would look like in an engineering procedure manual that has seven sections. Section One would have engineering procedure numbers from EP-01-01 through EP-01-99. This allows for 99 different procedures.

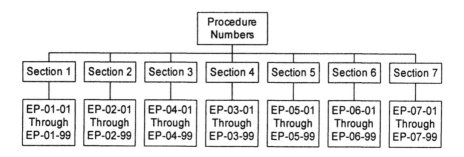

Figure 6.3. Engineering procedure numbers.

6.8.1   Example—Engineering Procedure Numbers and Interpretation

Following are some examples of how to identify the engineering procedure by its assigned number. The first two letters identify the department function that is responsible for the document, the next two digits identify the manual section, and the last two digits uniquely identify the procedure.

Complete Number	Interpretation
EP-05-10	This is the 10th procedure in manual Sec. 5.
EP-07-05	This is the 5th procedure in manual Sec. 7.

6.9.0   ENGINEERING PROCEDURE NUMBER ASSIGNMENT

Because the numbering system in this part of the book is for a manual paper-based documentation system, a logbook is required to record each engineering procedure number. When assigning new engineering procedure numbers, the question always is, what is the next available number? A logbook displays the last number that was assigned, and provides a place for a new number to be entered.

6.9.1   Engineering Procedure Number Assignment Log

Following is an example page from the document number assignment log for engineering procedures that is maintained by engineering document control. With this log you will be answering the following questions:

- What is the engineering procedure's number?
- What is the title of the engineering procedure?
- Who assigned the number?
- When was the number assigned?

![globe] Engineering Procedure Number Assignment Log			
EP No.	Title	Initials	Date
❶	❷	❸	❹

Example–Engineering Procedure Number Assignment Log Sheet

Engineering Procedure Number Assignment Log—Preparation. Each of the following circled numbers corresponds to the circled numbers on the example engineering procedure number assignment log.

❶ EP No.: Enter the engineering procedure's number.

❷ Title: Enter the engineering procedure's title.

❸ Initials: Enter your initials after logging in the information.

❹ Date: Enter the date after the information is loaded into the log.

6.9.2   Example—Engineering Procedure Number Assignment Log

Below is an example of a completed page from an engineering procedure number assignment log.

6.10.0   ENGINEERING PROCEDURE TITLES

The engineering procedure title needs to identify which function is being addressed in such detail that there is no question as to what it is covering.

Engineering procedure title examples would be:

Design Reviews

Part Numbering System

Engineering Change Request

Forms Control

🌐 Engineering Procedure Number Assignment Log			
EP No.	Title	Initials	Date
EP-01-01	Product Planning	PAC	7/22/00
EP-01-02	Product Introduction	PAC	8/12/00
EP-01-03	etc.		.
EP-01-04			
EP-01-05			
EP-01-06			
EP-01-07			
EP-01-08			
EP-01-09			
EP-01-10			
EP-01-11			
EP-01-12			
EP-01-13			
EP-01-14			
EP-01-15			
EP-01-16			
EP-01-17			
EP-01-18			
EP-01-19			
EP-01-20			
EP-01-21			
EP-01-22			

Example–Engineering Procedure Number Assignment Log (with sample entries)

6.11.0 FORM NUMBERS

All forms will need unique identification numbers assigned to them for tracking purposes (Fig. 6.4). Identification numbers are used as locators in spreadsheets and in engineering document control. There are two numbering systems shown in this book. This section is for a manual paper-based documentation system and Ch. 7 is for electronic.

Following is an example of an identification number for a form.

E001 = Form

E = One letter department designation for engineering.

001 = Three digit sequential logbook number (allows for 999 separate engineering form numbers).

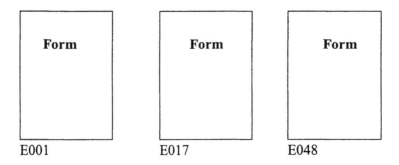

E001   E017   E048

Figure 6.4. Form number structure.

6.11.1 Example–Form Numbers and Interpretation

Following are some examples of how to identify the form by its assigned number. The letter identifies the department function that is responsible for the form, and the last three numbers uniquely identifies each separate form.

Complete Number	Interpretation
E019	This is the 19th engineering form.
E035	This is the 35th engineering form.

6.12.0   FORM NUMBER ASSIGNMENT

Because the numbering system in this part of the book is for a manual paper-based documentation system, a logbook is required to record each form number. When assigning new numbers, the question always is what is the next available number? A logbook displays the page that shows the last number that was assigned, next to it is a place for a new number, and so on.

6.12.1   Example—Form Number Assignment Log

Following are example pages from the document number assignment log for forms that is maintained by engineering document control. With this log you will be answering the following questions:

- What is the form's number?
- What is the title of the form?
- Who assigned the number?
- When was the number assigned?

Form Number Assignment Log—Preparation. Each of the following circled numbers corresponds to the circled numbers on the example form number assignment log.

➊ Form No.: Enter the form's number.

➋ Title: Enter the form's title.

➌ Initials: Enter your initials after logging in the information.

➍ Date: Enter the date after the information is loaded into the log.

![globe logo] Form Number Assignment Log			
Form No.	Title	Initials	Date
❶	❷	❸	❹

Example–Form Number Assignment Log Sheet

⊕	Form Number Assignment Log		
Form No.	Title	Initials	Date
E001	Document Review Form	PAC	7/12/00
E002	Document Change Request Form	PAC	8/19/00
E003	etc.		
E004			
E005			
E006			
E007			
E008			
E009			
E010			
E011			
E012			
E013			
E014			
E015			
E016			
E017			
E018			
E019			
E020			
E021			
E022			

Example–Form Number Assignment Log (with sample entries)

6.13.0 MANUAL NUMBERS

All engineering procedure manuals will have unique numbers assigned to them for identification and tracking purposes. Identification numbers are used as locators in spreadsheets and in engineering document control. Figure 6.5 illustrates the numbering system for engineering procedure manuals. There are two numbering systems shown in this book. This section is for a manual paper based documentation system and Ch. 7 is for electronic.

Following is an example of an engineering procedure manual's identification number.

EPM001

EPM = Three letter document designation for engineering procedure manual.

001 = Three digit sequential logbook number (allows or 999 separate engineering procedure manual numbers).

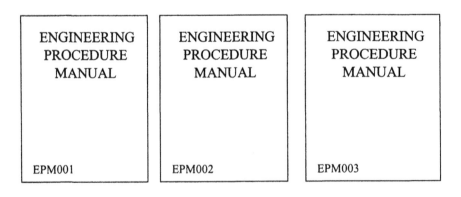

Figure 6.5. Engineering procedure manual number structure.

6.13.1 Manual Numbers and Interpretation

Following are some examples of how to identify a manual by its assigned number. The first three letters identify the department function that is responsible for the manual and the last three numbers uniquely identifies each separate manual.

Complete Number	Interpretation
EPM025	This is the 25th engineering procedure manual.
EPM068	This is the 68th engineering procedure manual.

6.14.0 MANUAL NUMBER ASSIGNMENT

Because the numbering system in this part of the book is for a manual paper-based documentation system, a logbook is required to record each manual number. When assigning new numbers, the question always is, what is the next available number? A logbook displays the last number that was assigned, provides a place for a new number, and so on.

6.14.1 Example—Manual Number Assignment Log

Following are example pages from the document number assignment log for manuals that is maintained by engineering document control. With this log you will be answering the following questions:

- What is the manual's number?
- What is the title of the manual?
- Who assigned the number?
- When was the number assigned?

(🌐) Manual Number Assignment Log			
Manual No.	Assigned To:	Initials	Date
❶	❷	❸	❹

Example–Manual Number Assignment Log Sheet

⊕ Manual Number Assignment Log		Initials	Date
Manual No.	Assigned To:	Initials	Date
EPM001		ABC	10/22/00
EPM002		DEF	11/12/00
EPM003			
EPM004			
EPM005			
EPM006			
EPM007			
EPM008			
EPM009			
EPM010			
EPM011			
EPM012			
EPM013			
EPM014			
EPM015			
EPM016			
EPM017			
EPM018			
EPM019			
EPM020			
EPM021			
EPM022			

Example–Manual Number Assignment Log (with sample entries)

Manual Number Assignment Log—Preparation. Each of the following circled numbers corresponds to the circled numbers on the example manual number assignment log.

❶ Manual No.: Enter the manual's number.

❷ Assigned to: Enter who the manual was assigned to.

❸ Initials: Enter your initials after logging in the information.

❹ Date: Enter the date after the information is loaded into the log.

6.15.0   DOCUMENT REVIEW FORM NUMBERS

Document Review forms will need unique identification numbers assigned to them for logging and tracking purposes. Identification numbers are used as locators in spreadsheets and in engineering document control. There are two numbering systems defined in this book. This section is for a manual paper-based documentation system and Ch. 7 is for electronic.

Following is an example of an identification number for a Document Review form.

DRF001 = Document Review form

DRF = Three letter document designation for Document
     Review form

001 = Three digit number (this allows for 999 Document
     Review forms.)

Figure 6.6 shows an example of what Document Review form numbers would look like on the form.

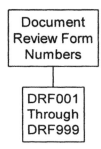

Figure 6.6. Document review form numbers.

6.15.1 Example—Document Review Form Numbers and Interpretation

Following are some examples of how to identify a Document Review form by its assigned number. The first three letters identify the type of document, and the next three digits uniquely identify the Document Review form.

Complete Number	Interpretation
DRF004	This is the 4th Document Review form prepared.
DRF015	This is the 15th Document Review form prepared.

6.16.0 DOCUMENT REVIEW FORM NUMBER ASSIGNMENT

Because the numbering system in this part of the book is for a manual paper-based documentation system, a logbook is required to record each Document Review form number. When assigning a new number, the question always is, what is the next available number? A logbook displays the last number that was assigned, and provides a place for a new number to be entered.

6.16.1   Document Review Form Number Assignment Log

Following is an example page from the document number assign-ment log for document review forms that is maintained by engineering document control. With this log you will be answering the following questions:

- What is the document review form's number?

- When is the document's number?

- What is the document's title?

- Who assigned the number?

- What date was the number assigned?

Document Review Form Number Assignment Log—Prepara-tion. Each of the following circled numbers corresponds to the circled numbers on the example, document review form number assignment log.

❶ DRF No.: Enter the document review form's number.

❷ Document No.: Enter the document's number.

❸ Title: Enter the document's title.

❹ Initials: Enter your initials after logging in the information.

❺ Date: Enter the date after the information is loaded into the log.

Following is an example of a completed page from a document review form number assignment logbook.

🌎 Document Review Form Number Assignment Log				
DRF No.	Document No.	Title	Initials	Date
❶	❷	❸	❹	❺

Example–Document Review Form Number Assignment Log Sheet

🌐 Document Review Form Number Assignment Log				
DRF No.	Document No.	Title	Initials	Date
DRF001	EP-05-02	Single Source Authorization	PAC	2/1/00
DRF002	EP-06-11	Bill of Material	PAC	2/14/00
DRF003	etc.			
DRF004				
DRF005				
DRF006				
DRF007				
DRF008				
DRF009				
DRF010				
DRF011				
DRF012				
DRF013				
DRF014				
DRF015				
DRF016				
DRF017				
DRF018				
DRF019				
DRF020				
DRF021				
DRF022				

Example–Document Review Form Number Assignment Log (with sample entries)

6.17.0 DOCUMENT CHANGE REQUEST FORM NUMBERS

Document Change Request forms will need unique identification numbers assigned to them for logging and tracking purposes. Identification numbers are used as locators in spreadsheets and in engineering document control. There are two numbering systems defined in this book. This section is for a manual paper-based documentation system and Ch. 7 is for electronic.

Following is an example of an identification number for a Document Change Request form.

DCRF001 = Document Change Request form

DCRF = Three letter document designation for Document Change Request form

001 = Three digit number (this allows for 999 Document Change Request forms.)

Figure 6.7 shows an example of what Document Change Request form numbers would look like on the form.

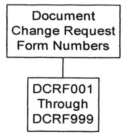

Figure 6.7. Document change request form numbers.

6.17.1   Example—Document Change Request Form Numbers and Interpretation

Following are some examples of how to identify a Document Change Request form by its assigned number. The first three letters identify the type of document, and the next three digits uniquely identify the Document Change Request form.

Complete Number	Interpretation
DCRF003	This is the 3rd Document Change Request form prepared
DCRF017	This is the 17th Document Change Request form prepared

6.18.0   DOCUMENT CHANGE REQUEST FORM NUMBER ASSIGNMENT

Because the numbering system in this part of the book is for a manual paper-based documentation system, a logbook is required to record each Document Change Request form number. When assigning a new number, the question always is, what is the next available number? A logbook displays the last number that was assigned, and provides a place for a new number to be entered.

6.18.1   Document Change Request Form Number Assignment Log

Following is an example page from the document number assignment log for document change request forms that is maintained by engineering document control. With this log you will be answering the following questions:

- What is the document change request form's number?
- What is the title of the document?
- Who assigned the number?
- When was the number assigned?

DRF No.	Document No.	Title	Initials	Date
❶	❷	❸	❹	❺

Document Change Request Form No. Assignment Log

Example–Document Change Request Form Number Assignment Log Sheet

Document Change Request Form Number Assignment Log—Preparation. Each of the following circled numbers corresponds to the circled numbers on the example document change request form number assignment log.

❶ DCRF No.: Enter the document change request form's number.

❷ Document No.: Enter the document's number.

❸ Title: Enter the document's title.

❹ Initials: Enter your initials after logging in the information.

❺ Date: Enter the date after the information is loaded into the log.

6.18.2 Example—Change Request Form Number Assignment Log

Below is a completed example page from a change request form number assignment logbook.

6.19.0 DOCUMENT MASTER LIST

You will need to establish a Document Master List of all required engineering documentation. Spreadsheets can be used for managing the list. The Document Master List will include policies, departmental instructions, engineering procedures, forms, manuals, document review forms, and document change request forms.

🌐 Document Change Request Form No. Assignment Log				
DCRF No.	Document No.	Title	Initials	Date
DCRF001	EP-01-09	Product Structuring	PAC	2/1/00
DCRF002	EP-03-05	Product Specification	PAC	2/14/00
DCRF003	etc.			
DCRF004				
DCRF005				
DCRF006				
DCRF007				
DCRF008				
DCRF009				
DCRF010				
DCRF011				
DCRF012				
DCRF013				
DCRF014				
DCRF015				
DCRF016				
DCRF017				
DCRF018				
DCRF019				
DCRF020				
DCRF021				
DCRF022				

Example–Document Change Request Form Number Assignment Log (with sample entries)

Following is an example page from the document master list that is maintained by engineering document control. With this list you will be answering the following questions:

- What is the document number?

- When is the revision of the document?

- What is the title of the document?

- How many pages are in the document?

- Who entered the information into the master list?

- When was the information entered?

Document Master List—Preparation. Each of the following circled numbers corresponds to the circled numbers on the example document master list.

❶ Document No.: Enter the document number.

❷ Rev.: Enter the document's revision letter.

❸ Document Title: Enter the document's title.

❹ Number of Pages: Enter the total number of pages.

❺ Initials: Enter your initials after the information is loaded into the log.

❻ Date: Enter the date after the information is loaded into the log.

6.19.1 Example—Document Master List

Below is an example page of a completed document master list.

🌐 Document Master List					
Document No.	Rev.	Title	Number of Pages	Initials	Date
❶	❷	❸	❹	❺	❻

Example–Document Master List Sheet

⊕ Document Master List					
Document No.	Rev.	Title	Number of Pages	Initials	Date
D-01		Departmental Instructions			
D-02					
D-03					
etc.					
DCRF001		Document Change Request Forms			
DCRF002					
DCRF003					
etc.					
E001		Form Numbers			
E002					
E003					
etc.					
EP-01-01		Engineering Procedures (Section 1)			
EP-01-02					
EP-01-03					
etc.					
EPM001		Engineering Procedures Manuals			
EPM002					
EPM003					
etc.					
P-01		Policies			
P-02					
P-03					
etc.					

Example–Document Master List (with sample entries)

7

Electronic Database

This chapter shows an electronic method of assigning document numbers and establishing an electronic database. The number assignment method detailed in Ch. 6 is for a manual paper-based documentation system. All documents will need to have unique numbers assigned to them for identification and tracking purposes. The document number is used as a locator (find number) in a database, and identifies each unique document type.

7.1.0   DOCUMENT NUMBER STRUCTURE

Because the database is structured in such a way that allows for searching for any of the document's information, the document number does not need to have very much intelligence in it. Use the document type for the prefix and then a two-digit number for the suffix.

Policy number example: P01

P   =   Prefix

01   =   Suffix

7.1.1   Prefix

The prefix part of the document number defines the document type. When you sort the document type column all of the different document types will be grouped together. For example, D would put all of the departmental instructions next to each other in a report.

P   =   Policy

D   =   Departmental Instructions

EP =   Engineering Procedure

7.1.2   Suffix

The suffix contains the number of documents of each type that is allowed. For example, 00 would allow for 99 different documents.

7.1.3   Identification Numbers and Interpretation

Complete Number	Interpretation
P16	This would be the 16th Policy ever written
D12	This would be the 12th Departmental Instruction ever written
EP09	This would be the 9th Engineering procedure ever written

Table 7.1 shows what the document database might look like over time.

Table 7.1. Database Report Printout

Document No.	Title
D01	Departmental Instructions
D02	
D03	
E01	Engineering Procedures (Sec. 1)
E02	
E03	
EP01	Policies
EP02	
EP03	
etc.	

7.2.0   DOCUMENT DATABASE

You will need to develop a database for policies, department instructions, engineering procedures, and their associated review and change documents. This is the key database that is used for planning, scheduling, and status reporting for all of the documents. The database is used for identifying the latest available document number for each type of document. Following is an example page from the database that is maintained by engineering document control. With this list you will be answering the following questions:

- What is the document number?

- When is the revision of the document?

- What is the title of the document?

- How many pages are in the document?

- What is the document review form number?

- What is the document change request form number?

- Who entered in the information into the database?

- When was the information entered?

Document Database—Preparation. Each of the following circled numbers corresponds to the circled numbers on the example document database.

❶ Document Number: Enter the document number.

❷ Revision: Enter the document's revision letter.

❸ Title: Enter the document's title.

❹ Number of Pages: Enter the total number of pages.

❺ DRF Number: Enter the document review form number.

❻ DCRF Number: Enter the document change request form number.

❼ Initials: Enter your initials after the information is loaded into the log.

❽ Date: Enter the date after the information is loaded into the log.

7.2.1 Example—Document Database

Following is an example page of a completed document database.

🌐	**Document Database**		
Document Number	❶		
Revision	❷		
Title	❸		
Number of Pages	❹		
DRF Number	❺		
DCRF Number	❻		
Initials	❼		
Date	❽		
etc.			

Example–Document Database Sheet

(🌎)	Document Database		
Document Number	D01	D02	D03
Revision	A	C	B
Title	Master File Update	Document Release	Document Number Assignment
Number of Pages	5	3	6
DRF Number	DRF005	DRF003	DRF010
DCRF Number	DCRF019	DCRF001	DCRF056
Initials	PAC	PAC	PAC
Date	03/22/01	05/01/01	O8/12/01
etc.			

Example–Document Database (with sample entries)

Appendix A: Policy Example

Included in this appendix is an example of a policy template that is filled in to show you how it is developed. Each of the document's elements is covered in the template.

See Ch. 3 for more examples of policies.

(🌐)	Policy	
Title: Engineering Lab Notebooks		Policy No.: P-02
Release Date: 5/3/00		Page: 1 of 2

1.0 Purpose

The purpose of this policy is to provide the guidance for the use of engineering lab notebooks to document potential patentable designs.

2.0 Scope

This policy applies to all personnel involved in the design of new products.

3.0 Policy

Engineering lab notebooks will serve as the initial documentation of the conception of inventions, as well as a daily record of technical investigations, analysis, and findings. Data recorded in engineering lab notebooks will form the basis for product design and it will establish a legal record of conception and witnessing of technical innovation. Specific requirements regarding engineering lab notebooks are set forth in an engineering procedure. Compliance is essential, as these notebooks could serve as evidence in a legal action arising from patent disputes.

All engineers involved in design and development activities will keep and maintain engineering lab notebooks. Each engineer will be responsible for the security of his or her notebooks. Engineering lab notebooks will be restricted and confidential and will be treated as such.

Prepared By:	
Engineer:	Date:

Approved By:	
Engineering Manager:	Date:
Vice President Engineering:	Date:

Example–Policy Template Page 1

(🌍)	Policy	
Title: Engineering Lab Notebooks		Policy No.: P-02
Release Date: 5/3/00		Page: 2 of 2

3.0   Policy (Cont'd.)

Engineers and technicians will be required to comply with this policy as a condition of employment. Engineering lab notebooks will be the property of the company and, as such, will be surrendered to engineering document control upon termination or job change.

Engineering lab notebooks will be supplied, issued, numbered, and controlled by engineering document control. The front cover of each notebook will have a non-removable label containing the following information: department, name of responsible engineer or technician, company name address, date allocated, and notebook control number.

Example–Policy Template Page 2

Appendix B: Departmental Instructions Example

I have included an actual departmental instruction template that is filled in to show you how it is developed. Each of the departmental instruction elements is covered in the template. I have also included an example of a Document Release Notice.

See Ch. 3 for more examples of departmental instructions.

(🌎)	**Departmental Instruction**	
Title: Document Release		Dept. Inst. No.: D-02
Release Date: 2/19/00		Page: 1 of 6

1.0   Purpose

The purpose of this procedure is to provide instructions, and assign responsibilities for controlling the release of product documentation. It will also provide the instructions for preparing and submitting the Document Release Notice form.

2.0   Definitions

Product Documentation: Documents that are prepared and released by Engineering for the purpose of defining and controlling the manufacture and inspection of company products. The documents that will need to be controlled are: specifications and drawings.

Release: The process of transferring custody of product documentation from the originator to engineering document control.

Document Release Notice Form: The Document Release Notice authorizes formal release and distribution of new and revised documents. Processed Document Release Notice forms are available for review in engineering document control.

Prepared By:	
Engineer:	Date:
Approved By:	
Engineering Manager:	Date:

Example–Department Instruction Template Page 1

🌎	**Departmental Instruction**	
Title: Document Release		Dept. Inst. No.: D-02
Release Date: 2/19/00		Page: 2 of 6

2.0   Definitions (Cont'd.)

Control of Documents: After documents are released, the originator relinquishes physical control and tracking responsibilities by submitting them to engineering document control for recording, copying, distributing, and filing.

Distribution: Copies of documents bearing the official stamp that identifies them as having been released. As revisions to documents are released, they will receive the same distribution as their last distribution, unless requested otherwise.

3.0   Document Release Procedure

This procedure is divided into the following parts:

- Document Release Notice Procedure
- How to Prepare the Document Release Notice form

The following procedure describes who is responsible and what they are supposed to do for each processing step.

3.1   Engineer

<u>Steps</u>

1. Prepare the Document Release Notice form to release new or revised documents into the documentation system.

Example–Departmental Instruction Page 2

(🌎)	**Departmental Instruction**	
Title: Document Release		Dept. Inst. No.: D-02
Release Date: 2/19/00		Page: 3 of 6

3.1   Engineer (Cont'd.)

Steps

2. Forward the completed Document Release Notice form to the Engineering Manager for review and approval.

3.2   Engineering Manager

3. Verify that the Document Release Notice form is complete and correct. If approved, sign and date, then return the form to the Engineer for further processing.

3.3   Engineer

4. Attach document originals to the Document Release Notice form then forward the package to engineering document control for further processing.

3.4   Engineering Document Control

5. Enter the next available number in the logbook and on the Document Release Notice form.

Example–Departmental Instruction Page 3

⊕	Departmental Instruction	
Title: Document Release		Dept. Inst. No.: D-02
Release Date: 2/19/00		Page: 4 of 6

3.4   Document Control (Cont'd.)

6.   Log the document information from the Document Release Notice form into the database.

7.   Run copies of the documents that are attached to the Document Release Notice form and red stamp them with the appropriate stamp and date. The stamp is determined by the type of release: Development, Prototype, or Production.

8.   Sign and date the form. This date is the official release of documents into the system.

9.   Forward a copy of the form to the Engineer.

10.   Attach copies of the released documents to the Document Release Notice form and forward it to either a or b:

a. Production Control for distribution to manufacturing. Down level copies of the documents shall be returned to engineering document control to be destroyed.

b. Purchasing for distribution to vendors, as necessary. Purchasing is responsible for notifying vendors about destroying copies of superseded documents.

11.   File the Document Release Notice form and the new or revised document originals that are attached to the form.

12.   Red stamp "OBSOLETE" on the down level originals that are on file, then date.

13.   Destroy all copies of the superseded documents that are on file throughout the company.

Example–Departmental Instruction Page 4

(🌐)	**Departmental Instruction**	
Title: Document Release		Dept. Inst. No.: D-02
Release Date: 2/19/00		Page: 5 of 6

5.0   Overview

The procedure for processing the Where-Used input form shall be followed by each individual responsible for entering information on the form. Each circled number below corresponds to the circled number on the following Where-Used Input form E023.

5.1   Engineer

❶ Enter your name and date that the Where-Used Input form was completed.

❷ Enter the next available sequential number.

5.2   Engineering Manager

❸ Enter your name and date if approved.

5.3   Engineer

❹ Mark the appropriate box:

Routine – will be processed within five days.

Urgent – will be processed the same day they are received.

❺ Mark the appropriate box. The type of release determines this.

❻ Enter the document number of each document being released. Attach a separate sheet, if releasing more documents than are allowed for on the form.

Enter the revision letter of each document being released, such as: "A", "B", etc. The release process does not cause the document's revision letter to advance.

Enter the title of each document being released.

⊕ **Departmental Instruction**	
Title: Document Release	Dept. Inst. No.: D-02
Release Date: 2/19/00	Page: 6 of 6

❻ (Cont'd.)

Enter the document letter size of each document being released, such as: A = (8½" × 11"), B = (11" × 17"), C = (17" × 24"), or D = (24" × 36").

Enter the number of pages for each document that is being released.

5.4   Engineering Document Control

❼ Sign and date the Document Release Notice form. This is the official release date of the documentation into the system.

Example–Departmental Instruction Page 6

Document Release Notice					
Engineer: ❶		Date:		DRN No.: ❷	
Engineering Manager: ❸ Date:				Priority: ❹ __Routine __Urgent	
__ Document Release: __ Development __ Preproduction __ Production ❺					
Documents To Be Released					
Document No. ❻	Revision	Document Title		Size	No. of Pages
Engineering Document Control: ❼			Release Date:		

Form E044 (Procedure EP-07-02) 6/15/00

Example–Document Release Notice

Appendix C: Engineering Procedure Example

I have included an actual engineering procedure template that is filled in to show you how it is developed. Each of the engineering procedure elements is covered in the template. I have also included an example of an Engineering Change Proposal.

See Ch. 3 for more examples of engineering procedures.

(🌐)	**Engineering Procedure**		
Title: Engineering Change Proposal		Eng. Proc. No.: EP-02-05	
Release Date: 03/15/00	Revision: B		Page: 1 of 5

1.0   Purpose

The purpose of this procedure is to provide instructions, and to assign responsibilities for preparing and submitting a proposed change to the customer for evaluation and disposition. It will also provide instructions for preparing the Engineering Change Proposal form. It will also provide instructions for preparing and submitting the Part Number request form.

2.0   Applicable Documents

Engineering Procedure:
    EP-2-2 Document Number Assignment Logbook

Engineering Form:
    E031 Engineering Change Proposal

Prepared By:	
Engineer:	Date:

Approved By:	
Engineering Manager:	Date:
Manufacturing Manager:	Date:
Quality Manager:	Date:

Example–Engineering Procedure Template Page 1

(🌐)	**Engineering Procedure**	
Title: Engineering Change Proposal		Eng. Proc. No.: EP-02-05
Release Date: 04/15/00	Revision: A	Page: 2 of 5

3.0   General Requirements

3.1   The Engineering Change Proposal is an engineering change package that consists of a completed form and other supplemental data, as necessary. An Engineering Change Proposal shall be provided for each proposed change which may result in a technical and/or hardware change subsequent to the establishment of an approved technical and/or hardware/software baseline.

3.2   All data elements shall be included in an Engineering Change Proposal with exception of an urgent Engineering Change Proposal. An entry shall be made in each line. Not applicable (N/A) will be used only after due consideration. The urgent Engineering Change Proposal may be utilized for all changes proposed prior to establishing the completed hardware/software baseline defined by the Configuration Baseline Document.

4.0   Engineering Change Proposal Procedure

This procedure is divided into the following parts:

• Engineering Change Proposal
• Engineering Change Proposal Form Preparation

The following procedure describes who is responsible and what they are supposed to do for each processing step.

Example–Engineering Procedure Page 2

⊕	**Engineering Procedure**	
Title: Engineering Change Proposal		Eng. Proc. No.: EP-02-05
Release Date: 04/15/00	Revision: A	Page: 3 of 5

4.1 Engineer

<u>Steps</u>

1. Prepare the Engineering Change Proposal form E031 to submit a proposed change to the customer for evaluation and disposition.

2. Forward the completed Engineering Change Proposal to the Engineering Manager for review and approval.

4.2 Engineering Manager

3. Verify that the Engineering Change Proposal form is complete and correct. If approved, sign and date, then return the form to the Engineer for further processing.

4.3 Engineer

4. Forward the approved Engineering Change Proposal Form to Engineering Document Control for the assignment of part numbers.

4.4 Engineering Document Control

5. Enter the next available number in the logbook and on the Engineering Change Proposal Form.

6. Forward a copy of the Engineering Change Proposal form to the Engineer.

✪	Engineering Procedure	
Title: Engineering Change Proposal		Eng. Proc. No.: EP-02-05
Release Date: 04/15/00	Revision: A	Page: 4 of 5

7. Run copies, stamp, distribute, and then file the original in the Engineering Change Proposal file by part number.

4.5   Engineer

8. After receipt of the Engineering Change Proposal Request form from the customer, it shall be forwarded to the originator for disposition.

5.0   Overview

The procedure for processing the Engineering Change Proposal form shall be followed by each individual responsible for entering information on the form. Each circled number below corresponds to the circled number on the following Part Number Request form E031.

5.1   Engineer

❶ Engineer: Enter your name and date that the Engineering Change Proposal form was completed.

5.2   Engineering Document Control

❷ Engineering Document Control: Assign the next available part number from the Document Assignment Logbook EP-2-2, under Paragraph 9.0. Then enter the number in the Part Number Request form E031.

Example–Engineering Procedure Page 4

⊛	**Engineering Procedure**	
Title: Engineering Change Proposal		Eng. Proc. No.: EP-02-05
Release Date: 04/15/00	Revision: A	Page: 5 of 5

5.3   **Engineering Manager**

❽ Engineering Manager: Enter your name and date upon approval of the information entered on the form.

5.4   **Engineer**

❹ Priority: Check the priority of the change. (Urgent is for line down situations only.)

❺ Model Number: Enter the model number(s) that are affected by this change.

❻ Serial Number Affectivity: Enter the serial number, part name and drawing number.

❼ Part Number: Enter the part number, part name and drawing number.

❽ Reason for Change: Enter the reason for the change.

❾ Description of Change: Enter the description of the change.

5.5   **Customer**

❿ Approved by: Enter your name and date, if approved.

Engineering Change Proposal		
Engineer: ❶	Date:	ECP No.: ❷
Engineering Manager: ❸ Date:		Priority: ❹ __Routine __Urgent
Model(s) affected: ❺		
Serial number effectivity: ❻		
Part No.: ❼	Part Name:	Drawing No.:
Reason for the change: ❽		
Description of the change: ❾		
Approved by: (Customer) Name: ❿ Title: Date:		

Form E031 (Procedure EP-4-2) 8/12/00

Example–Engineering Change Proposal

Appendix D:
Forms, Lists, and Logs

Controlled Document Agreement Form

Engineering Procedure Manual Distribution List

Document Review Form

Document Change Request Form

Policy Number Assignment Log

Departmental Instruction Number Assignment Log

Engineering Procedure Number Assignment Log

Form Number Assignment Log

Manual Number Assignment Log

Document Review Form Number Assignment Log

Document Change Request Form Number Assignment Log

Document Master List

Document Database

See Ch. 4 under "Controlled Document Agreement" for information on how to develop and use the controlled document agreement form.

CONTROLLED DOCUMENT AGREEMENT

The contents of the manual has been specifically organized to assist those individuals assigned and employed within the company to perform their functions in an acceptable manner and in concert with other functions. The manual has been developed strictly and solely for the purpose of and use by employees of the company, and any disclosure, plagiarizing, or other use made of the structuring or contents is prohibited.

I, the undersigned, to whom this manual is assigned, fully understand and agree that I will use it for its intended purpose, will maintain it in a current and reasonable condition, and, in the event of my termination, will be held accountable for it and stand ready to forfeit it when asked to do so.

Manual Number:_____

Signed:_____

Date:_____

Form E100 (Procedure EP-7-3) 7/10/00

Example–Controlled Document Agreement Form

See Ch. 4 under "Engineering Procedure Manual Numbers" for information on how to develop and use the engineering procedure manual distribution list.

🌐 Engineering Procedure Manual Distribution List			
Manual Number	Issued to:	Date Issued	Date Returned
EPM001			
EPM002			
EPM003			
EPM004			
EPM005			
EPM006			
EPM007			
EPM008			
EPM009			
EPM010			
EPM011			
EPM012			
EPM013			
EPM014			
EPM015			
EPM016			
EPM017			
EPM018			

Example–Engineering Procedure Manual Distribution List

See Ch. 5 under "Document Review Form Numbers" for information on how to develop and use the document review form.

(🌐)	**Document Review Form**	
Originator:	Date:	DRF No.:
Document Number:	Title:	

Documentation Types:

_____Policy _____Departmental Instruction
_____Engineering Procedure

Reviewer Response:

Mark your comments directly on the document. After your review is complete, either hold on to your mark-ups for the next Document Review Board meeting or return them to engineering. Engineering will bring all marked-up copies to the next meeting for discussion and agreement.

Your comments are due by:_____. If more time is needed, contact engineering for an extension.

_____ Approved

_____ Conditionally Approved (See attached comments.)

_____ Not Approved, explain.

Reviewer Signature: Date:

Distribution:

Engineering:

Manufacturing:

Quality:

Form E001 (Procedure EP-1-5) 7/12/01

Example–Document Review Form

See Ch. 5 under "Document Change Request Form Numbers" for information on how to develop and use the document change request form.

(🌐) **Document Change Request Form**		
Originator: Date:		DCRF No.:
Document Type: _____ Policy _____ Departmental Instruction _____ Engineering Procedure _____ Form	Disposition: ____Change ____Obsolete	
	Priority: ____Routine__Urgent	
Document Affected:		
Document Number	Revision	Title
Reason for change:		
Justification:		
Approved By:		
Engineering Manager:	Date:	
Manufacturing Manager:	Date:	
Quality Manager:	Date:	

Form E002 (Procedure EP-6-3) 8/19/01

Example–Document Change Request Form

See Ch. 6 under "Policy Numbers " for information on how to develop and use the policy number assignment log.

⊕ Policy Number Assignment Log			
Policy No.	Title	Initials	Date
P-01			
P-02			
P-03			
P-04			
P-05			
P-06			
P-07			
P-08			
P-09			
P-10			
P-11			
P-12			
P-13			
P-14			
P-15			
P-16			
P-17			
P-18			
P-19			
P-20			
P-21			

Example–Policy Number Assignment Log

See Ch. 6 under "Departmental Instruction Numbers" for information on how to develop and use the departmental instruction number assignment log.

🌐 Departmental Instruction Number Assignment Log			
Di No.	Title	Initials	Date
D-01			
D-02			
D-03			
D-04			
D-05			
D-06			
D-07			
D-08			
D-09			
D-10			
D-11			
D-12			
D-13			
D-14			
D-15			
D-16			
D-17			
D-18			
D-19			
D-20			
D-21			

Example–Departmental Instruction Number Assignment Log

See Ch. 6 under "Engineering Procedure Numbers" for information on how to develop and use the engineering procedure number assignment log.

🌐 Engineering Procedure Number Assignment Log			
EP No.	Title	Initials	Date
EP-01-01			
EP-01-02			
EP-01-03			
EP-01-04			
EP-01-05			
EP-01-06			
EP-01-07			
EP-01-08			
EP-01-09			
EP-01-10			
EP-01-11			
EP-01-12			
EP-01-13			
EP-01-14			
EP-01-15			
EP-01-16			
EP-01-17			
EP-01-18			
EP-01-19			
EP-01-20			
EP-01-21			

Example–Engineering Procedure Number Assignment Log

See Ch. 6 under "Form Numbers" for information on how to develop and use the form number assignment log.

(🌐)	Form Number Assignment Log		
Form No.	Title	Initials	Date
E001			
E002			
E003			
E004			
E005			
E006			
E007			
E008			
E009			
E010			
E011			
E012			
E013			
E014			
E015			
E016			
E017			
E018			
E019			
E020			
E021			

Example–Form Number Assignment Log

See Ch. 6 under "Manual Numbers" for information on how to develop and use the manual number assignment log.

⊕ Manual Number Assignment Log			
Manual No.	Assigned To:	Initials	Date
EPM001			
EPM002			
EPM003			
EPM004			
EPM005			
EPM006			
EPM007			
EPM008			
EPM009			
EPM010			
EPM011			
EPM012			
EPM013			
EPM014			
EPM015			
EPM016			
EPM017			
EPM018			
EPM019			
EPM020			
EPM021			

Example–Manual Number Assignment Log

See Ch. 6 under "Document Review Form Numbers" for information on how to develop and use the document review form.

🌐 Document Review Form Number Assignment Log				
DRF No.	Document No.	Title	Initials	Date
DRF001				
DRF002				
DRF003				
DRF004				
DRF005				
DRF006				
DRF007				
DRF008				
DRF009				
DRF010				
DRF011				
DRF012				
DRF013				
DRF014				
DRF015				
DRF016				
DRF017				
DRF018				
DRF019				
DRF020				
DRF021				
DRF022				

Example–Document Review Form Number Assignment Log

See Ch. 6 under "Document Change Request Form Numbers" for information on how to develop and use the document change request form.

🌐Document Change Request Form No. Assignment Log				
DCRF No.	Document No.	Title	Initials	Date
DCRF001				
DCRF002				
DCRF003				
DCRF004				
DCRF005				
DCRF006				
DCRF007				
DCRF008				
DCRF009				
DCRF010				
DCRF011				
DCRF012				
DCRF013				
DCRF014				
DCRF015				
DCRF016				
DCRF017				
DCRF018				
DCRF019				
DCRF020				
DCRF021				
DCRF022				

Example–Document Change Request Form Number Assignment Log

See Ch. 6 under "Document Master List" for information on how to develop and use the document master list.

🌐 **Document Master List**					
Document No.	Rev.	Title	Number of Pages	Initials	Date

Example–Document Master List

See Ch. 7 under "Document Database" for information on how to develop and use the document database.

🌐 Document Database			
Document Number			
Revision			
Title			
Number of Pages			
DRF Number			
DCRF Number			
Initials			
Date			
etc.			

Example–Document Database

Appendix E:
Further Reading

Cloud, P. A., (1997) *Engineering Procedures Handbook*, Noyes Publications, Westwood, NJ

Escoe, A., (1997) *Nimble Documentation: The Practical Guide for World-Class Organizations*; American Society for Quality

Hackos, J. T., (1994) *Managing Your Documentation Projects*, John Wiley & Sons

Hartman, P. J., (1999) *Starting a Documentation Group: A Hands-On Guide*, Clear Point Consultants, Inc.

Nagle, J. G., (1996) *Handbook for Preparing Engineering Documents: From Concept to Completion*, IEEE

Page, S. B., (1998) *Establishing A System of Policies and Procedures*, BookMasters, Inc.

WORLD WIDE WEB REFERENCES

www.stc-va.org —The Society of Technical Communicators

www.loc.gov —Library of Congress

www.amazon.com —Virtual book store

www.bn.com —Barnes and Nobel virtual book store

www.lehigh.edu/`ka3/sops/sop.index.html —Standard Operating Procedure

www.SOPS.com —Standard Operating Procedure Example

www.fsi.ncu.edu/acs/standards.htm —Standard and Procedures

www.zianet.com/abs/proc.HTM —Procedure Development

www.procedures.com —ISO 9000 Procedures

Index

Administrative
    costs   3
    procedures   20
Approved vendor list
    example form   55
Assigned number form
    examples   118
Assigning document numbers
    electronic method   139
Assignment log   163
    example   126
    policy number   105, 163
    policy number example   106

Back cover   63
Binders   63
Block number system   101
Blocks of numbers   101

Centralized procedure writing group
    functions   8
Change control   92
    process   87
Change request
    reject or accept   87

Changes incorporated   83
Collecting information   16
Complete engineering procedure   41
Controlled document agreement
    form   72, 163
Cover page   61
    example   71

Database   141
    example page   141
    structure   139
Departmental instruction number
    assignment   109
    assignment log   163
    assignment log example   111
Departmental instructions   20,
        108, 148
    develop a database   141
    example pages   30, 32, 34
    how many needed   20
    how to identify   109
    logbook   101
    questions   30
    required sections   34
    revision history list   96
    revision history sheet   93

template   148
title   111
Distribution copy   82
Distribution to departments
   memo example   72
Distribution to individuals
   memo example   72
Dividers   63
Document change request
   form   87, 88, 163
   example   90
   how to identify   132
   identification numbers   131
   number assignment log   163
   number assignment log
      example   134
   number logbook   132
   questions   88
Document control   2
   requirements   7
   system   6
Document database   140, 163
   example page   142
Document format standard   15
Document master list   101, 134, 163
   example page   136
Document number   139
   prefix   140
   suffix   140
Document number assignment
   log   132
   example page   104, 109,
      114, 123
   questions   128
Document review and approval
   method used   81
Document review board   10, 81,
      82, 92
   meeting   83
   memo   93
   review and approval   92
Document review flow   81
Document review
   form   82, 126, 163

example pages   85, 128
identification number   127
logbooks   101
number logbook   127
questions   83
Document review form number
   assignment log   163
Document titles   101
Document-numbering systems   101
Documentation
   approval   81
   components list   6
   review   81
   system   3, 6
Draft   82

Effective engineering program   59
Electronic based documentation
   system   126, 139
Electronic database   3, 139
Engineering department
   review and approval   82
Engineering document
   control   10, 83, 141
   identification number   118
Engineering documentation
   system   3
Engineering operation
   documentation   3
Engineering policies   18
Engineering procedure   1, 21, 30,
      54, 59
   associated documentation   4
   case study   4
   contents   45
   develop a database   141
   develop and manage   4
   developement   9
   example pages   41, 43, 46, 156
   format   41
   format and contents   20
   identification number
      example   113

ISO 9000 regulations 80
  logbooks 101
  management 9
  manual distribution 71
  revision history list 96
  revision history sheet 93
  template 156
Engineering procedure documentation
  organizing 8
  training or retraining 9
Engineering procedure manuals
  approving, publishing and
    distributing 59
  distribution list 163
  identification numbers 122
  numbering system 122
Engineering procedure number 113
  assignment log 163
  example 116
  logbook 114
Engineering procedure writing
    group 10
  reason for creating 11
Engineering systems
  policies 18

Find number 139
Form format 55
  questions 55
Form number
  logbook 119
Form number assignment log 163
  example page 119
Format and contents
  established 16
Forms
  documented evidence 54
  logbooks 101
  record 54
  unique identification number 118
Front cover 63

GMP. *See* Good manufacturing
    practices
Good manufacturing practices 3

House style 15

Identification number
  departmental instructions 108
Identifying number 72
Instructions
  step-by-step 21
Interviewing for information 16
ISO 9000 3, 20
  guidelines 3, 27, 34
  requirements 6

Key database 141
Key documents 7

Logbook 114

Manual copies
  control of 72
Manual distribution list 71, 76–78
Manual introduction and maintenance
    page 66
Manual number
  logbook 123
Manual pages
  engineering procedure
    example 64
Manual paper-based documentation
    system 108, 123, 126, 132, 139
  numbering system 109
Manual revisions 79

Manuals
  changes 59
  components 60
  distribution to departments 72
  distribution to individuals 72
  identification numbers 59
  logbooks 101
  reviewed 59
  three-ring binders 63
Marketing
  improvements 3
Master list 2
Memo example 79

Numbering system 104

Operating procedures
  write, change, and control 8
Operational efficiency 3

Paper-based system 3
Performing routine tasks
  goal 20
Plastic jackets 63
Policies 18
  develop a database 141
  example pages 145
  general record 17
  how many needed 18
  identification numbers 103
Policy
  example pages 24, 27
  format 23
  identify 104
  logbooks 101
  questions 24
  required sections 27
  revision history 93, 96
  template example 145
  title examples 106

Policy numbers 103
  assigning new numbers 104
  example 139
Prefix 140
Prioritize documents 14
Procedural information 59
Procedure manuals 5
Procedure writing group
  list of questions 14
Profitability 3

Quality documentation standard 20

Research and Development
  policy 19
Review board 92
Revision history list 96, 98
  example 100
Revision history sheet 93
  questions 93
Revision level 79

Sales activity
  improvement 3
Spine 62, 63
Suffix 140

Traditional documentation system 7

Write a policy 17
Writing group
  functions 14

Printed and bound by CPI Group (UK) Ltd, Croydon, CR0 4YY

03/10/2024

01040433-0010